Philosophy of Mathematics

Philosophy of Mathematics

Selected Writings

Charles S. Peirce

Edited by
Matthew E. Moore

INDIANA UNIVERSITY PRESS
Bloomington and Indianapolis

This book is a publication of

Indiana University Press
601 North Morton Street
Bloomington, Indiana 47404-3797 USA

www.iupress.indiana.edu

Telephone orders	800-842-6796
Fax orders	812-855-7931
Orders by e-mail	iuporder@indiana.edu

Manufactured in the United States of America

Cataloging information is available from the Library of Congress.

ISBN 978-0-253-35563-8 (cl.)   ISBN 978-0-253-22265-7 (pbk.)

1 2 3 4 5 15 14 13 12 11 10

Contents

Preface

The purpose of this book is to make Peirce's philosophy of mathematics more readily available to contemporary workers in the field, and to students of his thought. Peirce's philosophical writings on mathematics are reasonably well represented—despite the shortcomings detailed in Dauben (1996, 28–39)—in the *Collected Papers,* much more so in *The New Elements of Mathematics.* The Chronological Edition of Peirce's *Writings* will surely set the gold standards, in this area as in others, of comprehensiveness and textual scholarship; it will also provide the annotations and other auxiliary apparatus whose absence so greatly diminishes the usability of the earlier editions for all but the most seasoned Peirceans. But the Chronological Edition has only just reached the last quarter century of Peirce's life, when most of the central texts on mathematics (and nearly all of those in the present volume) were written. And its very comprehensiveness will make it difficult for those with a particular interest in the philosophy of mathematics to find their way to what they really need.

Hence this book. It does not pretend to be a comprehensive selection, even of Peirce's most important philosophical writings on mathematics. It seeks rather to be a selection of major texts that is comprehensive enough to serve as a serious introduction to his philosophy of mathematics, sufficient in itself for those whose primary interests lie elsewhere, and a stepping-stone for specialists to more advanced investigations. If it helps to turn some of its readers into specialists, so much the better; for the secondary literature on the mathematical aspects of Peirce's thought makes no more than a good beginning, and much remains to be done.

Precious little of that literature is due to philosophers of mathematics who come to Peirce for insight into the living problems of their discipline. I will argue in the introduction that some of Peirce's deepest insights into mathematics cannot be fully appreciated in isolation from his larger philosophical system (which is not to say that one must buy fully into the system in order to appropriate the insights). This creates something of a dilemma for a volume such as this one; for the system is complicated and unfinished. Anything like a thorough exposition would overwhelm the primary content, a redundancy for much of its intended audience, and a stumbling block for the rest. My imperfect solution has been to provide a very brief overview in the introduction, to be supplemented by more detailed piecemeal explanations in the headnotes to the individual selections. Those who feel the need of a more thorough systematic survey will do well to consult Nathan Houser's introductions to the two volumes of *The Essential Peirce,* and Cheryl Misak's to the *Cambridge Companion to Peirce.* These works provide convenient points of entry into Peirce's own writings and into the secondary literature.

An overlapping, and no less intractable, dilemma is posed by the explanatory apparatus: the endnotes and headnotes. As just noted, the currently available editions of these texts are distinctly lacking in such apparatus. The general aims are clear enough: the headnotes should prepare the reader for Peirce's own words by placing them in context, by outlining the argumentative structure, and by clearing away the most serious interpretive obstacles; the endnotes serve the same purposes, but on a smaller scale, smoothing out the residual bumps on what should already be a relatively well-marked road.

These generalities do not settle the practical question of what to include and what to omit. Even under the most favorable of circumstances this is something of a dilemma; for excess and defect alike diminish the value of a collection such as this. The dilemma is particularly acute with Peirce. He is an exceptionally allusive writer, whose reading was so wide that it is natural to despair, as with Pound's *Cantos,* of truly understanding him until one has done the impossible and read everything he did. Much of what he alludes to has now faded into obscurity; his mathematical references are often couched in the now unfamiliar technical language of his time, and even when updated will go over many readers' heads.

I have sought, as a rule, to provide substantive explanations of matters that are either (a) essential to a basic understanding of the text in question, and not common knowledge; or else (b) of the first importance for the reader who wants to go beyond a basic understanding. The second conjunct of disjunct (a) is meant to exclude anything that a reader with a standard philosophical education, and a serviceable grasp of college mathematics (including elementary calculus), can reasonably be expected to know: I presume, for example, that the reader knows who Descartes was and what the Pythagorean Theorem says. Disjunct (a) will as a rule evoke more detailed explanations than (b). In some cases (for example, the Gilbert quote in selection 20, where I couldn't resist), I have stretched (b) to the breaking point. Many apparent oversights in the apparatus reflect my (fallible) judgments of importance. There are also passages that have defied my best efforts to elucidate them; these I leave to subsequent inquirers. Any errors are all my own, but for the rest I am much indebted to those who have gone before, and especially to previous editions of Peirce's writings. It would badly clutter the apparatus of this volume to give separate acknowledgments of everything my annotations owe to earlier editors; I hope that this blanket acknowledgment will suffice.

Titles pose another problem for a collection that draws so heavily on unpublished manuscripts. My general rule of thumb has been to use Peirce's title where he provided one; where he did not, I have for the most part looked to the Robin catalog, or to the Chronological Edition. In a handful of cases I have supplied my own titles, despite my great reluctance to add to the cacophony. I have shortened a couple of Peirce's more prolix titles, and added bracketed annotations to a couple of the hopelessly uninformative ones. Any bracketed material in a title comes either from me or from one of the sources just mentioned.

Each headnote begins by naming the entry in the bibliography where a full citation to the text excerpted can be found. Works unpublished in Peirce's lifetime are cited by manuscript number in Robin's catalog. (For manuscripts reprinted in this book I give the call number at Houghton Library, the numerical portion of which is the same as the number assigned by Robin.) Citations to the standard editions of Peirce's writings consist of a volume number (where applicable) preceded by a single letter, with 'C' denoting the *Collected Papers,* 'N' *The New Elements of Mathematics,* 'E' *The Essential Peirce,* 'R' *Reasoning and the Logic of Things,* and 'W' the Chronological Edition of the *Writings;* following custom, I give paragraph numbers in citations to the *Collected Papers* and page numbers in the rest. Internal references, to selections in this book, will consist of a capital 'S' prefixed to a selection number and page number(s), all in parentheses: so, for example '(S1, 3)' directs the reader to selection 1, page 3. Page numbers alone in parentheses refer to the work (internal or external) most recently cited.

The bibliography itself has been divided into three parts, with primary texts in the first part, and the secondary literature divided between the second and third parts so that the second can serve as a guide to the literature on Peirce's philosophy of mathematics. The latter is selective, comprising works both known to me and, in my estimation, worth recommending. The endnotes, especially those for the introduction, provide further pointers into the bibliography, with the aim of helping the reader to find her way around in a body of scholarship that has already become quite extensive and in more than one place has taken on some depth. I hope that the reader can compensate for any congestion that results by skimming over the notes that provide too much information to suit her purposes.

This is not a critical edition, so there is very little comment on textual issues, except where I have had to make a decision that is obviously open to dispute. Though I have not updated Peirce's spelling or punctuation, I have silently corrected a handful of manifest errors in punctuation, and substituted 'and' for '&'. All other editorial substitutions and additions are placed in square brackets. In a few places, where there is potential confusion between authorial and editorial contributions, I have replaced Peirce's brackets with parentheses.

The organization of the book is roughly topical, beginning with selections on the nature of mathematics and its place in human knowledge (selections 1–4), followed by treatments of mathematical ontology and epistemology (5–10) and of the methods and objects of set theory (11–13), arithmetic (14–15), geometry (16–17), and the theory of continuity (18–29). The topical organization is only rough: as the reader will soon see, these texts show scant respect for the boundaries laid down by this or any other scheme. This is partly because Peirce's metaphysics and epistemology are so tightly integrated as to resist completely separate treatment, and partly because many of his larger ideas are most fully developed in connection with specific applications—for example, some of his most important ontological analyses of mathematics are found in the selections that focus on the theory of collections (sets). Within each topical group, the ordering of texts is chronological, except that the first selections on collection theory (11) and continuity (18) gather together short texts spanning several years and several stages, which are then fleshed out in the chronologically ordered texts that follow.

In order to keep the volume of manageable length, it has been necessary to limit its scope to the metaphysical and epistemological issues that have come to dominate the philosophy of mathematics as we know it. As a result there is nothing here about probability, a major preoccupation of Peirce's and an area in which his importance is widely acknowledged. Philosophers of mathematics have largely ceded probability to the philosophy of science, so there is some justification for this omission in the (arguably misdrawn) boundaries that divide these specializations. In any case Peirce's writings on probability deserve a volume like this one, all to themselves. Some readers may also feel that I have given unjustly short shrift to Peirce's many discussions of topology. Much of that material is more technical than philosophical, devoted largely to his reworking of Listing's Census Theorem. It is clear that Peirce thought this was of great importance, not least for his analysis of continuity; it is much less clear to me that his efforts in this direction proved as fruitful as he expected them to be. If I am wrong about that, then his work on topology, like that on probability, deserves a volume of its own, or at least a goodly share of a volume of writings on space and time, which certainly *do* deserve a volume of their own. No overview of Peirce's philosophy of mathematics can ignore topology altogether, but I have included only as much as I deem to be essential to such an overview. I have also omitted works—for example, Peirce's groundbreaking axiomatization of arithmetic (Peirce 1881)—whose indisputable importance for the foundations of mathematics are more on the technical than the philosophical side. The final category of omissions is the one that has cost me the most sleep: these are the cases where I could not convince myself that the benefits of inclusion were worth the exceptional cost, in sheer number of pages or in supplementary explanations. I can only hope that others will find my judgments on this score worth improving upon, and that the end result will be a clearer picture of what is really important in Peirce's philosophical writings on mathematics.

I initially assumed that it would be necesary to provide a good deal of information, both in the texts themselves and in the apparatus, on Peirce's technical work in logic. Certainly his discoveries in the logic of relations, and his development of the existential graphs, profoundly affected his thinking about mathematics and its methods. But as the book took shape I found, much to my surprise, that with a few notable exceptions,[1] Peirce's major philosophical writings on mathematics do not presuppose a knowledge of the technical details of his logical investigations. This is not to say that those details are irrelevant to his philosophy of mathematics; that is surely false, and I hope that this book will help to stimulate more advanced research along those lines. But since the goal of the book is to lay the groundwork for such inquiries, not to contribute to them, and since the cost (in pages and demands on the reader) would substantially outweigh the benefits to be gained by providing in-depth information on the logic, I have not provided it.

As with the acknowledgment of my debts to previous editors, avoidance of clutter has also prevented me from thanking every individual who has contributed to the apparatus at each point of contribution. So let me say here that where my own knowledge failed me I was fortunate to have the expert assistance of John Baldwin, David Blank, James Heitsch, Nick Huggett, Nadeem Hussain, Philip Kitcher, Michael Kremer, Kenneth Manders, Emily

Michael, Fred Michael, Angelica Nuzzo, Volker Peckhaus, Ahti-Veikko Pietarinen, John Skorupski, Gregory Sterling, Iakovos Vasiliou and Fernando Zalamea. I am especially grateful to those who provided much-needed technical and historical assistance with the mathematical background for the book: Philip Ehrlich, Robin Hartshorne, and Jérôme Havenel were more unstinting than I had any right to ask with their expert knowledge, respectively, of the mathematics of continuity, geometry, and topology. Randall Dipert's generous comments greatly improved the introduction. Joseph Dauben has come to my rescue time and again in matters pertaining to the history of mathematics in general, and to Peirce and Cantor in particular. I hope that the book reflects what I have learned from Dipert's and Dauben's writings on Peirce's mathematical philosophy.

As in all of my Peircean endeavors, I have leaned heavily on my friends at the Peirce Edition Project and the Institute for American Thought: Nathan Houser, André De Tienne, Cornelis de Waal, Jonathan Eller, and Albert Lewis. It is a pleasure to remember many long and fruitful conversations with Nathan and with André about Peirce's philosophy, and his thinking about mathematics. André has answered what must have seemed like an endless series of questions, ranging from handwriting to the curvature of space, with his characteristic good humor and his encyclopedic knowledge of Peirce. Having had the good fortune to work with such scholars, one comes to appreciate why Peirce so highly prized the community of inquirers.

Almost in tandem with this project, I have been involved with a forthcoming volume of new essays on Peirce's philosophy of mathematics, and have learned much from the contributions to that volume, whose authors (Daniel Campos, Elizabeth Cooke, Philip Ehrlich, Jérôme Havenel, Christopher Hookway, Susanna Marietti, Ahti-Veikko Pietarinen, Sun-Joo Shin, Claudine Tiercelin, Fernando Zalamea) are owed a special vote of thanks. The essays by Tiercelin and Hookway have been particularly important for me, because they are influenced by the problems and approaches in the philosophy of mathematics that have had the most influence on me: like Tiercelin, I have been much impressed by Penelope Maddy's objections to the Quine/Putnam indispensability arguments; and like Hookway, I am attracted to structuralism in mathematical ontology. The introduction to this volume is much indebted to them. Where there is any question about who got to an idea first, I am happy to concede priority, and likewise for the other authors just listed.

Putting this book together has given me a renewed appreciation of the obstacles that faced Peirce's early editors. It is nothing short of miraculous that Hartshorne and Weiss were able to fill in so much of the background to Peirce's work without the retrieval technologies that make scholarship so much easier for us. Among the many electronic resources that accelerated the completion of this book, I must make special mention of two. On innumerable occasions Google Books gave me ready access to books that would once have required a visit to a far-away library; and the St. Andrews MacTutor site, with their magnificent collection of capsule biographies of eminent mathematicians, was always my first stop in following up Peirce's references to his contemporaries and predecessors.

A number of individuals and institutions have helped to bring this book into being. This work was supported by two grants from the City University of New York PSC-CUNY Research Award Program. A Rodney G. Dennis Fellowship in the Study of Manuscripts, awarded by the Houghton Library at Harvard University, made it possible to spend most of the summer of 2008 in Cambridge working with the Peirce manuscripts in their collection. I am grateful to Houghton Library for permission to publish and quote from Peirce manuscripts or letters from its holdings. Thanks also to the Houghton staff for their patience and skillful assistance.

The staff of the Peirce Edition Project, who have already been thanked for their intellectual contributions, were just as indispensable on the practical side. I am particularly grateful to Albert Lewis and to Diana Reynolds for their invaluable assistance with software related matters, and to Cornelis de Waal and family for their hospitality. It has been a pleasure to work with Dee Mortensen, Michele Bird and their colleagues at Indiana University Press. Emily Michael deserves a special vote of thanks for making Brooklyn College such a hospitable environment for work on Peirce. Randall Dipert's generous assistance, in the initial phases, helped to get everything off to a good start. Joseph Dauben's ongoing advice and support have been crucial to this project, and to all my work on Peirce. Nathan Houser did so much to bring this book into being that it is hard to find the words to thank him. My debt to André De Tienne is also beyond reckoning. And as for Thomas Muller, all I can say to him is: you're the greatest.

Introduction

Charles Sanders Peirce's philosophy of mathematics plays a vital role in his mature philosophical system, and ought to play a more vital role in contemporary philosophical discussions of mathematics. The main business of this introduction will be to flesh out and defend these claims, and to exhibit their interdependence. The interdependence is important: the force of the claims is much diminished if we ignore it. We cannot fully appreciate the systematic importance of Peirce's philosophy of mathematics unless we appreciate its originality and depth *as* a philosophy of mathematics; and that originality and depth derive, in large part, from the distinctive systematic resources that Peirce can bring to bear in his philosophical analyses of mathematics.

At the same time, it must be admitted that linking the claims in this way makes it harder to defend them; for the system and the philosophy of mathematics are both in an unfinished state. A good deal of reconstruction is needed, and some new construction as well, before the full potential of either can be realized: far too large a job for a single book, let alone for the introduction to a single book. The point of walking you through the edifice is not to convince you to move in—it is not yet ready to be occupied—but rather to bring out the attractiveness of the overall design. This is not to say that a blueprint is all there is to show: a good deal of actual construction has been done, and some of the more finished portions of the structure are worthy of careful study, indeed of emulation. Even if you do not go so far as to pick up a hammer and help finish the job Peirce has started, you may pick up some ideas that will serve you well in your own philosophical constructions.

1. Mathematics in Peirce's Philosophy

Peirce was the second son of Benjamin Peirce, one of the foremost American mathematicians of his day. Young Charles received (arguably, suffered)[1] a rigorous mathematical training from his father, and though his own contributions to the subject were mainly in the area of logic, his writings clearly evince a professional's grasp of both recent and historical developments in a wide range of mathematical specialties.[2] One of the most daunting features of Peirce's philosophy for the beginner is his tendency to veer off without warning into complex mathematical arguments, whose bearing on the philosophical claims that precipitate them is not always easy to discern.

So Peirce was, like Quine's mathematical cyclist, a mathematical philosopher in a minimal conjunctive sense: he was both a philosopher and a mathematician. But he was a mathematical philosopher also in a deeper, less accidental, sense. He was a philosopher for whom mathematics was not just an important philosophical topic—though it certainly was that, for Peirce—but also a philosophical tool, which he turned to important philosophical purposes (Eisele 1979c). In 1894 he wrote to Francis Russell that it was his "special business to bring . . . modern mathematical exactitude into philosophy, and to apply the ideas of mathematics in philosophy" (Eisele 1979b, 277). The date is significant: it is near the beginning of the last quarter-century of Peirce's life, the period from which nearly every text in the present volume dates. This corresponds roughly to what Max Fisch calls the "*Monist* Period" of 1891–1914 (Fisch 1967, 192), the period in which Peirce arrived at what he himself called the "extreme realism"[3] of his later years. As his remark to Russell indicates, Peirce regarded his later system as a *mathematical* philosophy, built on the results and techniques of modern mathematics. Putnam and Ketner go so far as to suggest that the Cambridge Conferences Lectures ought to be entitled *The Consequences of Mathematics* because "Peirce's philosophy is a consequence of his mathematics" (R.2). It is no accident, then, that his most searching philosophical investigations of mathematics begin as the *Monist* period does; and it is with good reason that Murphey prefaces his treatment of Peirce's later philosophy with an extended "mathematical interlude" (Murphey 1961, 183–288). In a moment we will review some of the ways in which Peirce's philosophy came to be a mathematical philosophy. But first a few words about realism are in order.

One could (as Fisch more or less does in the paper just cited) trace Peirce's overall philosophical development through the evolution of what he means when he calls himself—as he steadfastly does from 1868 on (Fisch 1967, 187)—a "realist." Correlated with this terminological evolution is another, in the meaning of 'nominalism,' which Peirce uses (again, steadfastly after 1868) as a name for the deep-seated philosophical errors to which his realism is supposed to be the antidote. Throughout this extended development, 'realism' and 'nominalism' denote approximately and in part the opposed views on universals that the corresponding Latin terms denoted in the Middle Ages. But for Peirce there is a great deal more to the opposition than that: 'realism' serves at the same time as a general label for the theories of knowledge and reality that he puts forth as correctives to the errors of pre-decessors like Descartes and Hume.[4] The "extreme realism" of Peirce's *Monist* period commits him to the reality of possibilia, concreta, and laws. I will argue in §2 that these commitments have a great deal to do with Peirce's distinctive contributions to the philosophy of mathematics.

Now back to the respects in which Peirce's philosophy came to be, in these later years, a mathematical one. I will confine myself here to brief treatments of two of the most important mathematical aspects of Peirce's later thought: his grounding of philosophy in mathematics, and his philo-sophical deployment of mathematical ideas.

In saying that Peirce grounded philosophy in mathematics, I do not mean to suggest that he viewed philosophy as a branch of mathematics, or that he somehow reduced philosophical reasoning to mathematical reason-ing. For Peirce

> philosophy . . . is really an experimental science, resting on that experience which is common to us all; so that its principal reasonings are not mathe-matically necessary at all, but are only necessary in the sense that all the world knows beyond all doubt those truths upon which philosophy is founded. This is why the mathematician holds the reasoning of the meta-physician in supreme contempt, while he himself, when he ventures into philosophy, is apt to reason fantastically and not solidly, because he does not recognize that he is upon ground where elaborate deduction is of no more avail than it is in chemistry or biology. (S3, 20)

For Peirce philosophy rests on mathematics, not because it *is* mathematics, but because "logic ought to draw upon mathematics for control of disputed principles, and . . . ontological philosophy ought in like manner to draw upon logic" (S2, 13).[5] So he writes in "The Regenerated Logic," where he then goes on to explain these dependencies with reference to a Comtean classification of the sciences, wherein "each science draws regulating princi-ples from those superior to it in abstractness." Seven years later, in his Har-vard lectures on pragmatism, he puts forth a more complicated scheme in which the "normative sciences" (of which logic is one, though not the most fundamental) are held to be based upon "the science of phenomenology." Yet the ultimate dependence upon mathematics remains: "phenomenology does not depend upon any other *positive science* . . . [but it] must, if it is to be properly grounded, be made to depend upon the Conditional or Hypothet-ical Science of *Pure Mathematics*" (Peirce 1903e, E2.144). Phenomenology and the normative sciences (other than logic) are somewhat peripheral to our present concerns, so I will not delve more deeply here into these Peircean pronouncements. The pronouncements themselves should suffice to prove my present point: that in order to understand Peirce's later philosophy one must understand his philosophy of mathematics because mathematics lies, according to Peirce, at the base of the whole system of sciences and hence underlies philosophy.

Let us turn now to Peirce's philosophical deployment of mathematical ideas. His conception of the continuum is the shining example: it is the center of attention in many of our selections, and lies just under the surface of many others. As Peirce himself took pains to stress, the word 'continuum' is not being used here to refer to the object—the unique complete ordered field—that students of set theory and real analysis refer to by that name. Indeed, it is misleading to call the continuum as Peirce understood it an object at all: it would be better to call it a *kind* of object, and better still to find a word that does not obscure, as 'object' does, the radically *potential* nature of Peirce's continuum.[6] Some of the texts in this volume will cast light on these dark sayings, and also on the central place Peirce's continuum occupies in his mature philosophical system. Peirce went so far, in the eighth of the Cambridge Conferences Lectures, as to identify as "the [defining] characteristic of [his] doctrine . . . [his insistence] upon Continuity"; that is why, he says, he prefers "to call [his] theory Synechism, because it rests upon the study of Continuity" (Peirce 1992, R.261). In the passage just quoted Peirce identifies continuity with Thirdness, linking his theory of continuity directly to his triad of categories (more on those in §2). The Cambridge Conferences Lectures are rife with such linkages between continuity and Peirce's major philosophical concerns: "Generality," he says in the fifth lecture, "is logically the same as continuity" (Peirce 1992, R.190), a theme he sounds again in selection 12, from three years later; and in the third he suggests that "the question of nominalism and realism has taken this shape: are any continua real?" (S22, 175).

It is clear then, in a general way, that Peirce expects his continuum to do serious philosophical work. But we have not yet seen how he applies his *mathematics* of the continuum to the solution of philosophical problems. The selections in this book include two classic applications of this sort: the consciousness of time, and fallibilism.[7] First, some mathematical preliminaries.

Two distinctive features of Peirce's mathematical conception of the continuum emerge clearly in the relevant texts in this volume, and set it apart from what he calls the "pseudo-continuum" of Cantor, Dedekind, and what is now the standard presentation of the calculus. The first is that "in a continuous expanse, say a continuous line, there are continuous lines infinitely short" (S20, 156). Peirce's theory of the continuum thus agrees with more recent theories, like Robinson's nonstandard analysis, that affirm the existence of infinitesimal (infinitely small) quantities.[8] Indeed, one notable reconstruction of Peirce's continuum—Putnam's, included in the introduction (37–54) to Peirce (1992)—makes much of the parallels between Peirce and Robinson. But we must not make *too* much of them, for reasons connected with the second of our two features of the Peircean continuum.

In the manuscript of selection 20 Peirce states this feature in a sentence inserted, apparently as an afterthought, directly after the one just quoted. But it is in this apparent afterthought that he departs most radically from what has become the standard view of the continuum. For it is now usual to think of the geometric line in analytic terms, as a set of points. Peirce's insertion is a rejection of this analytic view: "In fact," he writes, "the whole line is made up of such infinitesimal parts." Putting this together with what precedes it we see that the "parts" in question are not points but rather "continuous lines infinitely short." This is what Fernando Zalamea calls the *reflexivity* of Peirce's continuum. In selection 25 Peirce attributes the idea to Kant, who "always defines a continuum as that of which every part . . . has itself parts" (204). Since unextended points have no parts, this immediately implies that lines are not composed of points, as the analytic view would have it. If the radicalism of the conception has not yet come across, consider the following claim, which Peirce regularly conjoins with his rejection of the analytic view: "on any line whatever . . . there is room for any multitude of points however great" (S18, 138). On the standard conception of the line, it is composed of points which can be put into one-one correspondence with the real numbers. Peirce does not maintain (as a partisan of nonstandard analysis might) that we need a grid with more points than that in order to fill up the line; he maintains that *no* grid, no matter how many points it contains, can fill up the line: "breaking grains of sand more and more will only make the sand more broken. It will not weld the grains into unbroken continuity" (139).

So how does all of this play out philosophically? "The Law of Mind" (Peirce 1892), the source of selection 19, is a treasure trove of philosophical applications of Peirce's continuum, so let us start there. One of the fundamental applications comes early in the paper, in the section entitled "Continuity of Ideas." Peirce argues there that "consciousness must essentially cover an interval of time; for if it did not, we could gain no knowledge of time, and not merely no veracious cognition of it, but no conception whatever" (145). However, this interval through which we are conscious cannot be finite in length; for if it were, then each of us would be immediately aware of *all* the past contents of her consciousness:

> If the sensation that precedes the present by half a second were still immediately before me, then, on the same principle, the sensation immediately preceding that would be immediately present, and so on *ad infinitum.* Now, since there is a time, say a year, at the end of which an idea is no longer *ipso facto* present, it follows that this is true of any finite interval, however short.

Peirce takes the only way out: "We are forced to say that we are immediately conscious through an infinitesimal interval of time." It is crucial to this account of our consciousness of time that a continuum should contain infinitely small intervals, as on Peirce's theory of the continuum it does.

The second application is less straightforward. Peirce makes use of his continuum in developing his fallibilism, the view that in our knowledge "exactitude, certitude and universality are not to be attained" (Peirce 1893d, C1.142).[9] A few pages later he writes that

> the principle of continuity is the idea of fallibilism objectified. For fallibilism is the doctrine that our knowledge is never absolute but always swims as it were in a continuum of uncertainty and of indeterminacy. Now the doctrine of continuity is that *all things* so swim in continua. (S20, 158)

The "as it were" here might lead one to worry that Peirce has been betrayed, by a picturesque expression of fallibilism, into the mistaken belief that he has found an illuminating connection between fallibilism and continuity. What is needed, if we are to allay the worry, is a non-metaphorical explanation of precisely how our knowledge "swims . . . in a continuum of uncertainty."

Peirce does not give us such an explanation in selection 20, but he does give us some hints there and elsewhere about how to move beyond the metaphor. He makes the reasonable-sounding suggestion that if all things are continuous, then physical measurement can never be completely accurate: "where there is continuity the exact measurement of real quantities is too obviously impossible" (Peirce 1893d, C1.172). More adventurously, he writes in his definition of 'synechism' for Baldwin's *Dictionary* that a synechist, that is, one who "insists upon . . . the necessity of hypotheses involving continuity" (Peirce 1901, C6.169), would never be satisfied with the hypothesis that matter is composed of atoms, all spherical and exactly alike. . . . [Neither] the eternity of the atoms nor their precise resemblance is, in the synechist's view, an element of the hypothesis that is even admissible hypothetically. For that would be to attempt to explain the phenomena by means of an absolute inexplicability. (173)

Here Peirce refuses to regard atoms—and the allegedly ultimate and inexplicable hypothesis about their nature—as analogous to points, which have no parts, no further structure that remains to be understood. Indeed, in his definition of 'synechism' he says that "the general motive [of synechism] is to avoid the hypothesis that this or that is inexplicable" (171). Ultimate explanations, which are themselves inexplicable, would be discontinuities (conceptual unextended points, as it were) and as a result untrue to the continuous reality they purport to explain.[10] In the same vein, Peirce charges the scientific infallibilist precisely with being blinded by "discontinuous" theories:

The ordinary scientific infallibilist . . . is committed to discontinuity in regard to all those things which he fancies he has exactly ascertained, and especially in regard to that part of his knowledge which he fancies he has exactly ascertained to be *certain.* . . . Thus scientific infallibilism draws down a veil before the eyes which prevents the evidences of continuity from being discerned. (S20, 158)

There is still a good deal of metaphor here, even when we have restored the argument in the second ellipsis. But the general connection between synechism and fallibilism is clear: because reality is continuous, our theories and explanations can never be more than approximately correct, and can never attain such a pitch of precision and completeness that nothing remains to be explained.

So far this sounds like an epistemological limitation, which it is, but that is not the whole story: Peirce also forges a metaphysical connection between synechism and fallibilism. Here is how he continues the passage just quoted:

But as soon as a man is fully impressed with the fact that absolute exactitude never can be known, he naturally asks whether there are any facts to show that hard discrete exactitude really exists. That suggestion lifts the edge of that curtain and he begins to see the clear daylight shining in from behind it. (158)

This suggestion of an ineliminable inexactness, not just in our knowledge, but in the very nature of things, forms the bridge in selection 20 between Peirce's discussion of continuity and his discussion of evolution. He goes on to argue (in a section of the manuscript omitted from this volume) that the diversification that evolution brings about would be impossible if the laws of nature were exceptionless; for in that case there would be no room for the spontaneity that gives rise to diversification: "mechanical law can never produce diversification" (Peirce 1893d, C1.174). Moreover, the inexactness of natural laws makes possible an evolutionary explanation of those laws themselves:

> Once you have embraced the principle of continuity no kind of explanation will satisfy you except that they grew. . . . [Laws] at any rate being absolute could not grow. They either always were, or they sprang instantaneously into being like the drill of a company of soldiers. This makes the laws of nature absolutely blind and inexplicable. . . . The fallibilist . . . asks may these *forces* of nature not be somehow amenable to reason? May they not have naturally grown up? After all, there is no reason to think they are absolute. If all things are continuous, the universe must be undergoing a continuous growth from existence to non-existence. (175)

Here Peirce draws together a number of the overarching themes of his later philosophy: fallibilism, tychism ("the doctrine that absolute chance is a factor of the universe" (Peirce 1898b, R.260)), and his evolutionary cosmology. His theory of continuity lies, as we have just seen, at the heart of this complex of ideas.

Peirce had good reason, then, to accept 'synechism' as a label for his later philosophical system. Clearly we cannot understand—let alone assess—that system unless we understand his theory of continuity. Murray Murphey makes this point forcefully at the end of his book on Peirce's philosophical development, and concludes that "the grand design" of Peirce's later system "was never fulfilled" because

> Peirce was never able to find a way to utilize the continuum concept effectively. The magnificent synthesis which the theory of continuity seemed to promise somehow always eluded him, and the shining vision of the great system always remained a castle in the air. (Murphey 1961, 407)

Others (for instance, Putnam and Zalamea) have been more optimistic about the prospects for a reconstruction of the Peircean continuum.[11] The question remains open: there is much to be said on both sides of it. In any case it is clear that the answer will profoundly influence our assessment of Peirce's philosophical achievement.

2. CONTRIBUTIONS FOR TODAY

It is often said that Peirce anticipated this or that subsequent development in philosophy, or logic, or mathematics, or even in some field more remote from his main areas of concern (Eisele 1979d). Such claims can be of interest to the historian, if Peirce played an unrecognized causal role in those subsequent developments. But they are not enough, in and of themselves, to convince contemporary workers in the relevant fields that they ought to study Peirce more carefully. If he merely anticipates what those workers already know, then they have nothing to gain from reading him: they can applaud his prescience, and move on. So if we as philosophers of mathematics do have something to gain, as I maintain we do, from reading Peirce, there must be important respects in which he does *not* just anticipate us, some approaches he takes to his problems that can give us a new bearing on our own.[12]

There are general reasons to expect, even without looking in detail at Peirce's philosophy of mathematics, that he can offer us such new perspectives on the issues we find pressing. On the one hand, he was recognizably concerned with the basic philosophical questions about mathematics that we still grapple with—more on that in a moment. He was moreover an active member of the international community of mathematicians and philosophers who set much of the agenda, and laid much of the conceptual groundwork, for the philosophy of mathematics as know it. Peirce's contributions in logic had if anything a greater impact than those of other, now more widely studied, members of that community, including Frege—see Dipert (1995, 41–45), and also Putnam (1982). In his lifelong concern with "the logic of science," and in his sophisticated philosophical analyses of language, he is akin to the pioneers of the analytic tradition in philosophy; in many ways he speaks, if not the very language of analytic philosophy, at least a dialect that an analytic philosopher can recognize and come to master.

On the other hand, Peirce was in many respects outside what turned into the philosophical mainstream.[13] This is due partly to his relative isolation, after his dismissal from Johns Hopkins in 1884, from the community of professional researchers. But only partly. Peirce's philosophical education and temperament were in many ways different from those of the founders of the analytic tradition. Like Frege, he both revered and quarrelled with Kant; but Peirce's knowledge of Kant was more thorough, as was his knowledge of the history of philosophy, and the struggle with Kant shapes Peirce's philosophy in a more fundamental way than it does Frege's. Like Russell and the early positivists, Peirce was consciously influenced by the British empiricists, but he had at the same time a more acute awareness than the positivists did of the empiricist tradition's philosophical dead ends. Perhaps the most striking differences between Peirce and the analytic mainstream are his creative appropriation of scholastic realism, and his very circumscribed and circumspect, but nonetheless genuine, respect for Hegel.[14]

Peirce shares, then, many of our concerns; but his philosophical toolkit, though it has much in common with ours, is stocked with some unfamiliar equipment as well. These are the general reasons I alluded to for expecting that Peirce will have some surprising and worthwhile things to say to a philosopher of mathematics in our own day. In arguing more specifically for Peirce's ongoing relevance it will be necessary to look more closely at some of the less familiar equipment in his kit, and in particular to give some account of his wider philosophical system. But I want to begin, not there, but rather with some perennial problems of the philosophy of mathematics.

Those problems can be summed up with a variation on a serious joke of Russell's: Mathematics may be defined as the subject in which we never know what we are talking about, nor how we know that what we are saying is true.[15] Any adequate, unified philosophy of mathematics must simultaneously answer both Russell's metaphysical question (what is the subject matter of mathematics?) and his epistemological one (how do we acquire and justify our knowledge of that subject matter?). Philosophical discussions of mathematics have been overshadowed, since the publication of Benacerraf's "Mathematical Truth" (Benacerraf 1973), by doubts about the very possibility of such a unified account. Benacerraf shows that the *prima facie* best answers to Russell's questions (taken singly) are apparently incompatible. As metaphysicians, we will be drawn to the conclusion that mathematics is about abstract (non-physical, non-mental) objects; and as epistemologists we will be drawn to the view that our knowledge of any object begins in causal interactions with it. Benacerraf's Dilemma now immediately ensues: since abstract objects cannot enter into causal interactions, how can we ever come to know anything about them? Benacerraf's original formulation of his dilemma presupposed a causal theory of knowledge, which has subsequently fallen from favor. But his argument can be recast so that the Dilemma highlights the apparently insurmountable challenge that mathematical knowledge poses for any empiricist epistemology.[16]

One way to avoid this dilemma altogether is to deny that mathematical statements are true. Perhaps the most prominent contemporary advocate of this approach is Hartry Field, who advances a fictionalist view of mathematics on which its assertions are, like those composing a fictional narrative, truth-valued and false (Field 1980). A fictionalist is under no obligation to explain how we come to know about abstract objects; for if fictionalism is correct, mathematics gives us no reason to think that there are such things. It should be clear by now that a straightforwardly fictionalist philosophy of mathematics is not an option for Peirce. As we have seen, mathematics is on the ground floor of his hierarchy of the sciences; and the first few selections in this volume are particularly rife with references to mathematics as a science.[17] This is not yet enough, of course, for Peirce to fall afoul of Benacerraf. One can blunt the metaphysical horn of the Dilemma by scaling back the mathematics, and its ontological commitments, in the service of epistemic tractability; or one can blunt the epistemological horn by being more generous than an austere empiricism in estimating our capacities for knowledge of abstracta.

When all is said and done, Peirce's philosophy of mathematics deploys both of these strategies (though not, of course, with an eye to evading a dilemma stated several decades after his death). But his deployments are extremely subtle, and it is this subtlety, which derives in large part from the sophistication and originality of his wider philosophical system, that makes his philosophy of mathematics so worth our while. At first glance, though, it appears that Peirce simply casts himself on both horns at once. On the metaphysical side, he shows no inclination to cut back on the mathematics whose knowability he will ultimately need to account for. In selection 22 he commits himself, in effect, to a countably infinite sequence of uncountably infinite cardinals (171). Granted, this falls short of the theories of infinity that have come down to us from Cantor, but Peirce's ontological stinginess, relative to what has become the tradition, arises from what we would now regard as technical mistakes and not from any principled *horror infiniti.*[18] Peirce is thus, for all intents and purposes, a *mathematical realist,* that is, one who takes the bulk of classical mathematics to be, not just truth-valued, but true.

He is moreover, in some sense of the term, an empiricist: in the second of his Harvard Pragmatism Lectures (*HPL,* from here on in) he declares that "Experience is our only teacher" (*HPL,* E2.153); and the first of the "cotary propositions" he states in the seventh (*HPL,* E2.226) is that "*Nihil est in intellectus quin prius fuerit in sensu*" (nothing is in the intellect which was not prevously in the senses). It seems, then, that Peirce opts, in metaphysics and in epistemology alike, for the *prima facie* best answers that generate Benacerraf's Dilemma.

We can begin to locate Peirce's position by noting that, and why, he would reject a couple of well known ways out of the Dilemma. One way out, originating with Frege, is to class mathematical knowledge with logical knowledge. Though Frege himself was no empiricist, philosophers of that bent have been attracted to variations on his logicism because logical knowledge can plausibly be held to involve no abstract objects and hence to pose no problem for an empiricist epistemology. When Peirce considers the reduction of mathematics to logic, it is Dedekind rather than Frege whom he mentions; but what he says on the subject leaves little doubt that he would have rejected Frege's logicism as well. The roots of Peirce's anti-logicism run deep, and we will have occasion to unearth a number of them as we go along.[19] We have already turned up one, in his declaration that "logic ought to draw upon mathematics for control of disputed principles"; as the classification of sciences sketched on page xvii above makes clear, we ought to say that logic is (a branch of) mathematics and not the other way around.[20]

Quine's Indispensability Argument offers the empirically minded realist another way around Benacerraf's Dilemma. No developed branch of natural science can get by without mathematics; this is Quine's Indispensability Thesis. Combined with a holistic account of confirmation, this implies that empirical evidence confirms those mathematical statements that cannot be eliminated from our best scientific theories. So our empirical evidence for those theories is also empirical evidence for mathematical realism. Now add in Quine's Indispensability Criterion for ontological commitment: this says that the acceptance of a theory commits us to the existence of those entities that must be in the range of the theory's quantifiers in order for the statements composing the theory to be true. The Indispensability Thesis takes in existentially quantified sentences of the form "there is a function f such that . . ." and so our best scientific theories commit us to the existence of functions; and similarly for numbers, sets and the rest. So the Indispensability Argument underwrites, not just mathematical realism, but also platonism, the claim that mathematical truths are about abstract objects.[21]

There is much in this argument, and in Quine's general outlook, which Peirce could applaud. They are at one in their rejection of Cartesian foundationalism, among other things. But Peirce would also have a good deal of sympathy for two of the main objections to the Quinean approach, and his brand of mathematical realism may point towards a congenial alternative for those who are impressed by those objections.

The first objection is that Quine

> leaves unaccounted for precisely the *obviousness* of elementary mathematics. . . . [There are] very general principles that are universally regarded as obvious, where on an empiricist view one would expect them to be bold hypotheses, about which a prudent scientist would maintain reserve, keeping in mind that experience might not bear them out. (Parsons 1979–1980, 101–102)

Peirce, too, takes the obviousness of mathematics very seriously. To see just how seriously, let us delve a little further into his anti-logicism. Peirce goes so far as to deny, not just that mathematics *is* logic, but that it even *needs* logic: "in the perspicuous and absolutely cogent reasonings of mathematicians . . . appeals [to logic] are altogether unnecessary" (S1, 8). He does not deny that mathematicians make mistakes in reasoning; in selection 4 (31) he claims only a "practical infallibility" for the results of such reasoning. What he does deny is that logic is necessary, or even particularly useful, for the correction of such mistakes when they do occur. We get a hint of this explanation in the just-quoted characterization of the "reasonings of mathematics" as "perspicuous and absolutely cogent." If mathematical reasoning is in some sense maximally perspicuous, then no other kind of reasoning can be capable, by dint of its greater perspicuity, of rescuing the mathematician when she goes wrong. Indeed, if one is struggling even with mathematical reasoning, which is as perspicuous as reasoning can get, she is unlikely to do better with any other kind of reasoning. Peirce makes this last point explicitly about logic in selection 4 (24): "if the mathematician ever hesitates or errs in his reasoning, logic cannot come to his aid. He would be far more liable to commit similar as well as other errors there." Hence, since logic *is* less perspicuous than mathematics—as Peirce repeatedly avers—it cannot serve as a guide, or even as a corrective, to mathematical reasoning.

Even if we accept the line of argument just sketched, and agree that the maximal perspicuity of mathematical reasoning implies the logician's uselessness to the mathematician, we need not accept the uselessness until the perspicuity has been established. Peirce's description of mathematical reasoning in selection 4 (27) begins to fill this gap:

> Suppose a state of things of a perfectly definite, general description . . . [and] suppose further, that this description refers to nothing occult,—noth-

ing that cannot be summoned up fully into the imagination. Assume, then, a range of possibilities equally definite and equally subject to the imagination. . . . [The] question whether in such a state of things, a certain other similarly definite state of things, could or could not, in the assumed range of possibility, ever occur, would be one in reference to which one of the two answers *Yes* and *No* would be true, but never both. But all the pertinent facts would be within the beck and call of the imagination; and consequently nothing but the operation of thought would be necessary to render the true answer.

This passage is part of an explanation of the necessity of mathematical results, which Peirce traces to the fact (acknowledged by "all modern mathematicians") that "mathematics deals exclusively with hypothetical states of things, and asserts no matter of fact whatever." Putting necessity to one side for the moment, we can begin to see how the maximal perspicuity of mathematics falls out of Peirce's understanding of its aims and processes.

Just after noting the mathematician's exclusive concern with hypothetical states of things, Peirce goes on to say that "this is the true essence of mathematics," and the reader will find that he constantly affirms that the mathematician studies not what is, but what would be under a given hypothesis. In the long passage just quoted, Peirce begins to explain how the mathematician finds out about the hypothetical states of things that are her special province. Though many important details remain to be worked out, it is clear that both the hypothesis, and the further state of things whose compatibility with the hypothesis we wish to determine, must admit of a completely transparent representation, a representation fully "within the beck and call of the imagination." The transparency of the representation ensures that "nothing but the operation of thought [is] necessary to render the true answer."

We have now uncovered two grounds for the maximal perspicuity of mathematical reasoning. The first arises from its aim, which relieves the mathematician of any responsibility to the facts: there is no possibility here, as there is in every other science, of getting the facts wrong despite our best efforts (for instance, because through sheer bad luck our samples have been unrepresentative)—for the simple reason that the assessment of the mathematician's results does not involve the comparison of those results with the facts. The second source of perspicuity is the transparent and imaginatively tractable representations with which mathematical reasoning operates. Let us now consider these representations more closely, and the nature of the mathematician's operations on them. This will round out our initial discussion of

Peirce's epistemology for mathematics, and serve as a jumping-off point into his account of mathematical ontology.

In selection 7 (46) Peirce characterizes the mathematician's hypotheses as "always the conception of a system of relations," and then continues:

> In order that they may be reasoned about mathematically, these relations must be conceived as embodied in some kind of objects; but the character of the objects, apart from the relations, is utterly immaterial. They are always made as bare, skeleton-like, or diagrammatic as possible. With mathematicians not born blind, they are always visual objects of the simplest kind, such as dots, or lines, or letters, and the like.

The obvious paradigm is a geometrical diagram, the sort of picture one might draw when looking for a proof of a geometrical theorem, or when explaining one's proof to someone else. The skeletal simplicity of a diagram lends some plausibility to Peirce's claims about imaginative tractability: if all mathematical reasoning is based on diagrams of this sort, then its perspicuity is easily accounted for. In that case we can also see why "only blundering can introduce error into mathematics" (S5, 37), and why logical theory is of so little use in correcting such errors as do occur: what is more to the point is an injunction to look again, more carefully.

But *is* all mathematical reasoning based on diagrams of this sort? Even in geometry the tendency has been, since the rise of formalization, to downplay the role of diagrams (except, of course, for heuristic purposes). And pictorial diagrams like those in geometry play hardly any role at all in arithmetical and algebraic reasoning, as Peirce himself admits. Yet he steadily maintains that "the very life of mathematical thinking consists in making experiments upon diagrams and the like and in observing the results" (S6, 40). The phrase 'and the like' in this sweeping generalization suggests that Peirce's definition of 'diagram' is a broad one, and not restricted to the geometer's labelled pictures. In selection 7 we find a sketch of just such a broad definition, one that explicitly includes the "diagrams" of algebra:

> The diagrams in which the [mathematician's] hypotheses are embodied are of two kinds. In the one kind the parts of the diagram are seen in the visual image to have the relations supposed. In the other kind of diagrams, the parts have shapes to which conventions or "rules" are attached, by means of which the supposed relations are attributed, or imputed, to the parts of the diagrams. Geometrical figures are diagrams of the inherential kind, while algebraical formulae are diagrams of the imputations kind. (46)

With this sketch in hand, we can see how algebraic reasoning might fit into Peirce's sweeping claim about the "very life of mathematical thinking." An array of equations can plausibly be regarded as a diagram, as a visual depiction (given certain interpretive conventions) of a "system of relations"; and the rule-governed transformation of one array into another can plausibly be regarded as an experiment whose observable result is another diagram depicting (again, by way of the interpretive conventions) another system of relations which is a necessary consequence of the first. (In selection 10 (81), Peirce says that "*all* diagrams . . . depend upon conventions" [my emphasis]. This will turn out to be of great importance later on.)

We should take Peirce at his word, then, when he says that mathematical reasoning is diagrammatic. There is an important overlap here between Peirce's logic and his philosophy of mathematics. It is well known that one of his pioneering contributions to logic was the graphical system—encompassing not just propositional and first-order logic, but second-order and modal logic as well—of "Existential Graphs."[22] This system was, like Peirce's developed philosophy of mathematics, a product of his later years; it clearly influenced his views about the nature of mathematical reasoning, and in particular about its *diagrammatic* nature, and by his own account "was invented for the purpose of representing the reasonings of mathematics in as analytical a form as possible" (Peirce 1903g, N3.349).

Peirce's diagrammatic account of mathematical reasoning provides a hospitable setting for one of the more widely discussed ideas in his philosophy of mathematics: the distinction between theorematic and corollarial reasoning. Here is one of his more abstract explanations of the distinction:

> Any *Corollary* . . . would be a proposition deduced directly from propositions already established without the use of any other constructions than one necessarily suggested in apprehending the enunciation of the proposition Any *Theorem* would be a proposition . . . capable of demonstration from propositions previously established, but not without imagining something more than what the condition supposes to exist; and any such proposition would be a Theorem. (S8, 63)

The example with which Peirce goes on to illustrate these definitions makes the connection with diagrams explicit. He notes that the equality of the base angles of an isosceles triangle (the so-called *Pons Asinorum*)

> may be proved by first proving that a rigid triangle may be exactly superposed on the isosceles triangle, and that it may be turned over and reapplied

to the same triangle. But since the enunciation of the *Pons* says nothing about such a thing; and since the *Pons* cannot be demonstrated without some such hypothesis . . . it is a theorem. (63)

Note the *manipulation* of the diagram that is involved in the proof of Euclid's theorem. The example suggests that corollaries are results that can be read more or less directly off of the diagram of the hypothesis, while a theorem requires us to act upon the diagram in some way. In particular we must perform *new constructions* beyond those involved in the diagram of the theorem hypothesis. In the case of the *Pons* the latter diagram depicts a single isosceles triangle; in the proof we construct a new triangle, which is superposed on the first. Peirce claims that any proof of the result will require such a construction.[23] Moreover, if theorematic reasoning is essentially non-mechanical—as Peirce may or may not have believed: see Hookway (1985, 199–200)—the theorematic/corollarial distinction may help clarify the informal character of (some of) our mathematical knowledge, and thus draw some of the sting from Gödel's incompleteness theorems.

Our first divergence between Peirce and Quine, over the obviousness of mathematics, has taken us rather far into Peirce's account of mathematical knowledge. The second will do likewise for his account of mathematical existence. Quine started out as an anti-platonist (Goodman and Quine 1947), and never renounced the ontological minimalism that led him to take up that position; for him such minimalism was a guiding principle of scientific practice and thus also of a naturalistic philosophy like his, which sought to take the methods of science as its own. As a result, his ontology for mathematics is no more ample than it has to be in order to meet the needs of physical science. But mathematicians—set theorists particularly—develop theories whose apparent ontologies far exceed those needs. It seems that a minimalist must either advocate methodological reform, or else rehabilitate the offending theories by interpreting them in such a way that their offenses are only apparent. But to be a naturalist is to swear off of such philosophical carping at the established methods of science.

Peirce would surely agree with Penelope Maddy (1997, 158–160), who originated this objection to the Indispensability Argument, that the problem runs deep: the Quinean naturalist is not just coming up with the wrong answer, but is asking the wrong question to begin with.[24] Take, for example, one of the chief points of disagreement between Quine and most set theorists: the existence of a nonconstructible set. Quine favors Gödel's Axiom of Constructibility, which denies the existence of such sets, on minimalist grounds. For Quine the (non-)existence of a nonconstructible set is completely on a par with that of, say, electrons: the same kind of evidence is used in settling both questions, and 'exist' is univocal between the answers. Quine is implacably opposed to any attempt to distinguish different ways of being, in order to allow that questions of mathematical existence have a different sense, and are to be settled on different grounds, than other questions of scientific ontology.[25] For him, the set theorist who asserts the existence of a nonconstructible set is no less and no differently answerable to the facts than the physicist who asserts the existence of electrons.

This last way of putting it is calculated to hint at the fundamental disagreement here between Peirce and Quine. Peirce makes much of the mathematician's *freedom* from the facts. To quarrel with the set theorists about whether nonconstructible sets exist as electrons do is to fall prey to a profound misunderstanding of what they are up to: it is to lose sight of the fact that "mathematics deals exclusively with hypothetical states of things, and asserts no matter of fact whatever." This looks like the beginning of an attractive diagnosis of the difficulty with the Quinean approach, but it is only a beginning. We are owed an explanation of what hypothetical states of things are, one that is compatible both with the refusal to fight with the set theorists and with Peirce's insistence on mathematical truth and objectivity. Peirce is going to need an account of hypothetical states of things that is robust, but not *too* robust.

It is hard to see how one could give a decently robust account without *ipso facto* adopting some kind of modal realism; and that is in fact a component of the "extreme realism" outlined in §1 (p. xvii). At this point we can no longer defer the daunting task of filling in the outline, and giving a fuller picture of Peirce's wider philosophical system. A complete picture is of course out of the question; nor will it be possible to do justice, as any comprehensive account must do, to the development of Peirce's views. In default of attempting that, I will draw my very partial exposition mainly from *HPL,* one of Peirce's most synoptic accounts of his mature philosophy. Any such snapshot of so dynamic a thinker is bound to be imperfect; some of the selections in this volume can function as a supplement, and a corrective.[26]

The cornerstone of Peirce's metaphysics is his triadic system of categories. Though his labels for the categories took a while to settle down, as he recounts in selection 22, Peirce ultimately opted for the colorless numerical scheme of First, Second and Third. In *HPL* Peirce argues that the categories are "the three irreducible and only constituents of thought" (*HPL,* E2.165) and furthermore that "all three have their place among the realities of nature and constitute all there is in nature" (*HPL,* E2.178). His arguments are numerous and subtle; those for the former claim are largely phenomenological, and those for the latter proceed mainly by drawing out what Peirce takes to be the metaphysical implications of modern science. An important impetus to the scheme, which makes itself felt at several points in the lectures (e.g., at *HPL,* E2.170–173), is found in the notations of Peirce's logic of relatives, which (like the present-day notations that descend from them) represent propositions by means of *n*-place predicate symbols with "blanks" for the propositions' logical subjects. The three categories are then suggested by monadic, dyadic, and triadic predicates.[27]

Here is Peirce's summary explanation, which opens the third Harvard lecture, of his categories:

> Category the First is the idea of that which is such as it is regardless of anything else. That is to say, it is a *Quality* of Feeling.
>
> Category the Second is the Idea of that which is such as it is as being Second to some First, regardless of anything else and in particular regardless of any *law,* although it may conform to a law. That is to say, it is *Reaction* as an element of the Phenomenon.
>
> Category the Third is the Idea of that which is such as it is as being a Third, or Medium, between a Second and its First. That is to say, it is *Representation* as an element of the Phenomenon. (*HPL,* E2.160)

Peirce's illustration of Firstness, from the second lecture, is

> a consciousness in which there is no comparison, no relation, no recognized multiplicity . . . no change . . . nothing but a simple positive character. . . . Such a consciousness might be just an odor, say a smell of attar; or it might be one infinite dead ache; it might be the hearing of [a] piercing eternal whistle. (*HPL,* E2.150)

Peirce's modal realism is constituted in part by his commitment to the reality of Firsts; for as he explains in the syllabus to the Lowell Lectures, a quality "is . . . in itself, a mere possibility. . . . Possibility, the mode of being of Firstness, is the embryo of being. It is not nothing. It is not existence" (Peirce 1903h, E2.268–269).[28] The last sentence just quoted sets up the contrast between Firstness and Secondness, which *is* the category of existence in the strict Peircean sense of the term. Peirce typically illustrates Secondness by means of two-sided phenomena of action and reaction, effort and resistance, such as putting one's shoulder to a partly open but obstructed door (*HPL,* E2.150). Secondness is the category of actuality, of physical objects in causal interaction. Thirdness, finally, takes in all of those realities involving mediation; the sign relation, according to Peirce's famously triadic conception thereof, is an extremely important example, which is why in this summary he calls Thirdness "*Representation* as an element of the Phenomenon." But Thirdness is also the category of generality, and of law; one of the reasons Peirce came to lay such stress on continuity is precisely that he came to identify it with true generality, and hence to see an adequate understanding of continuity as essential to an adequate understanding of Thirdness. (Selection 22 contains one of Peirce's most searching explorations of continuity and generality.)

As just noted, Peirce's semiotics (theory of signs) has at its heart a triadic relation, namely that wherein a sign mediates between its object and its interpretant. A rattling sound in the brush, the word 'rattlesnake' whispered in your ear, the picture of a rattlesnake in my trail guide—all these Peirce would recognize as signs. The object of them all is the rattlesnake that the sound of our footsteps has put audibly onto the defensive. The interpretant is your thought of that object, so named because it interprets the sign. The three signs just mentioned exemplify three large classes of signs defined, in line with Peirce's categories, in terms of the relationship between sign and object. The picture in the trail guide functions as a sign of the snake by means of a resemblance between sign and object. Resemblance is constituted by shared qualities or Firsts, and a sign so constituted Peirce calls an *icon.* The sound coming from the brush is causally related to its object, namely, the snake that produces the sound by shaking its tail; a sign thus constituted by Secondness is called an *index.* A word like 'rattlesnake', by contrast, is connected to its object by means of a law (Thirdness), in this case a social convention of English usage; Peirce calls such a sign a *symbol.* Many readers will know that the icon/index/symbol trichotomy is but the tip of a massive taxonomical iceberg, comprising first ten and later sixty-six classes of sign. These readers will appreciate how much I have simplified even this one trichotomy.[29]

The third and final component of Peirce's system that I want to introduce, his pragmatism, is also the best known. In *HPL* (E2.134–135) he states the pragmatic maxim as

> the principle that every theoretical judgment expressible in a sentence in the indicative mood is a confused form of thought whose only meaning, if it has any, lies in a tendency to enforce a corresponding practical maxim

expressible as a conditional sentence having its apodosis in the imperative mood.

This is immediately followed, in the lecture, by his original formulation from "How to Make Our Ideas Clear" (Peirce 1878, E1.132): "Consider what effects, which might conceivably have practical bearings, we conceive the object of our conception to have. Then, our conception of these effects is the whole of our conception of the object." I presume that the general idea of pragmatism is familiar enough; so I am going to take that for granted and say a few words about Peirce's specific version of it. He came to see his first formulation as insufficiently realistic, owing to its refusal to consider a diamond to be hard if no attempt is ever made to scratch it (Peirce 1878, E1.132–133). A properly realistic pragmatism so explicates 'hard' that even an untouched diamond is hard because it *would* resist scratching if the attempt *were* made (Peirce 1905b, E2.356–357). The conviction that there is a fact of the matter about what the diamond would do under counterfactual conditions is an example of what Peirce is insisting on when he insists on the reality of Thirdness. A law, like that which governs the obdurate behavior of the diamond, is a Third, "whose Being consists in active power to establish connections between different objects" (Peirce 1908d, E2.435).[30] A First, say a simple quality, is a "may be," a way things *could* be; a Third is a "would be" or "conditional necessity" which determines how things *must* be whenever certain conditions are fulfilled. What differentiates Peirce's "extreme" scholastic realism from "the halting realism of Scotus" (Peirce 1905a, C6.175) is his commitment to these two kinds of generality (Boler 1963, 63–65, 148–149; Mayorga 2007, 136–141): not just the "may be"s that Scotus came close to acknowledging with his realism about universals, but also the "would be"s of conditional necessity (Thirdness, law). Note that this two-fold realism about "generals" is at the same time a two-fold modal realism.

Now let us see how Peirce puts these systematic resources to work in his mathematical ontology, and how nicely the result dovetails with his epistemology. Selection 10 has been much discussed in this connection, especially by Christopher Hookway, who has argued that in that text Peirce puts forth a variety of what is nowadays known as mathematical structuralism.[31] A structuralist maintains that the subject matter of a branch of mathematics is not a particular abstract object (or system of such objects) but rather a structure which many systems can have in common. So on a structuralist reading of number theory, for example, the number theorist studies the natural number structure rather than some privileged instance theoreof (say, some set of von Neumann ordinals). I agree with Hookway's interpretation, though I will not offer a detailed defense of it here: look to Hookway (forthcoming) for that, and for a careful placement of Peirce's position within the larger landscape of contemporary structuralism.[32] It should soon be evident that in selection 10 Peirce does portray mathematics as the "science of structure" and that his philosophical system enables him to give a distinctive and attractive answer to the vexed question of what mathematical structures are.

In that selection Peirce likens chemical "experimentation [which is] the putting of questions to nature" to the mathematician's "experiments upon diagrams . . . [in which] questions [are] put to the nature of the relations concerned" (80). This provokes the objection that "there is a good deal of difference between experiments like the chemist's, which are trials made upon the very substance whose behavior is in question, and experiments made upon diagrams, these latter having no physical connection with the things they represent." Peirce's answer to this objection goes right to the heart of his mathematical ontology. The chemist does indeed experiment "upon the very object of investigation," but that object is not, as the objector supposes, the individual sample that the chemist manipulates, but rather the "Molecular Structure, which in all his samples has as complete an identity as it is in the nature of Molecular Structure ever to possess" (81). So likewise the mathematician, in "experiments made upon diagrams," operates directly on the object of her inquiry, which is "the *form of a relation* . . . the very form of the relation between the two corresponding parts of the diagram."

Peirce turns the tables on his interlocutor by taking the Aristotelian position that the object of scientific knowledge, whether mathematical or physical, is not the concrete thing with which the scientist interacts, but rather the form that comes to be known as the result of the interaction. The Aristotelian tenor of the text is further reinforced by Peirce's choice of words: he has his objector say that the chemist experiments on a *substance,* but in his reply he speaks instead of the *object* of investigation, which turns out to something very much like an Aristotelian form, immanent in the substances the investigator operates upon. Once we have caught on to these Aristotelian overtones, we can hardly fail to hear as well the deliberate echoes of scholastic realism. Peirce says that in the chemist's various samples the molecular structure "has *as complete an identity as it is in the nature of Molecular Structure ever to possess.*" This qualification of his earlier affirmations of the sameness of the form is obviously derived from the medieval attempts to avoid the Boethian paradoxes that beset crude understandings of the identity of universals.

Peirce's admiration of Duns Scotus is well known, and Scotus is famous for his subtle solution to this very problem of identity. John Boler (1963) and, more recently, Rosa Mayorga (2007) have shown that Peirce's realism about "generals" can be fruitfully approached by way of his debts to, and departures from, Scotus's realism about universals. Peirce's conception of mathematical objects is a special case of this general rule. As the aside about identity reveals, Peirce does not regard the mathematician's forms as abstract individuals in a Platonic heaven; similarly Scotus argues that the Common Nature, the objective ground of a true predication, is not a Platonic form but has only a less than numerical yet real unity. As Boler (45–46) points out, Peirce agrees with Scotus not only in the kind of solution he offers to the problem of universals, but also in his understanding of the problem itself: for both it is the problem of "real commonness," of whether our general concepts have any basis outside the mind in something really common to the diverse things that fall under them. When we carry this over to the question of *mathematical* realism, we get something very much like Kreisel's contention that "the question of realism . . . is the question of the objectivity of mathematics and not the question of the existence of mathematical objects."[33] I believe that the selections in this volume will bear out the contention that Peirce's metaphysical analysis of mathematics is in line with that formulation of the question.

In an early declaration of his own scholastic realism Peirce praises Scotus for being "separated from nominalism only by the division of a hair" (Peirce 1871, E1.87).[34] The Common Nature that grounds a Scotistic universal in extramental reality has a kind of mind-dependence: its distinction from the nature that individuates the concrete thing is merely formal, consisting in the fact that "one, before the operation of the intellect, is conceivable without the [other] though inseparable from [it] even by divine power" (Grajewski 1944, 93). The objective grounding of a Scotistic universal is thus "separated by the division of a hair" from complete dependence on the mind; it is constituted by what we might call an objective conceivability, a permanent possibility of conception that inheres in things prior to the mind's operations. Peirce often refers to mathematical objects as *entia rationis* (beings of reason), but always takes care to dull the nominalistic edge of that expression with more realistic language, often in the very same breath. For example, in selection 12 he writes that "a *collection* is an *ens rationis*" and then immediately adds that "that reason or *ratio* that creates it may be among the realities of the universe" (97).

This comes at the tail end of an outline of Peirce's doctrine of "substantive possibility," a systematic elaboration of his realism about Firsts. The being of a quality is held to "[consist] in such logical possibility as there may be that a definite predicate should be true of a single subject" (96); the being of a collection consists in turn in the being of the quality that defines it.[35] This is one of many twists in Peirce's fascinating and unfinished theory of collections. It is also a particularly forthright expression of what we have already identified as a leading idea of his general ontology for mathematics: that the mathematician studies not what is, but what could be. So if Peirce is indeed a kind of structuralist, he can usefully be classed as a kind of *modal* structuralist, one for whom possible structures are the subject matter of mathematics.[36] Add to this his realism about modality, and you get a kind of mathematical realism, in Kreisel's mode: Peirce's commitment to the objective reality of Firsts induces a commitment to the objectivity of mathematics, which he so unwaveringly affirms.

In selection 13, Peirce says that "a collection is an abstraction, or is like an abstraction in being an *ens rationis.*" This mention of abstraction is a mere aside, but in selection 9 he analyzes not just collections, but also some fundamental geometricalia, as abstractions; and in selection 15 he does the same for the natural numbers, which on his account are abstractions from the practice of counting.[37] Indeed, in selection 9 the question of the reality of mathematical objects boils down to the question of the reality of abstractions. It is not altogether clear whether Peirce (should have) distinguished between abstractions and *entia rationis*.[38] But the connection is so close that it will do no harm to ignore the distinction here:

> An *ens rationis* may be defined as a subject whose being consists in a Secondness, or fact, concerning something else. (S13, 101)

> An abstraction is a substance whose being consists in the truth of some proposition concerning a more primary substance. (S9, 73)

The relevant sense of 'abstraction' may be somewhat unfamiliar to readers making their first acquaintance with these texts. When we hear the word, we typically think of what Peirce calls "precisive abstraction," that is, "that operation of the mind by which we pay attention to one feature of a percept to the disregard of others" (S4, 29). But what matters most for mathematical ontology is "hypostatic abstraction," which consists in taking a feature of a percept or percepts, after it has already been prescinded from the other elements of the percept, so as to take propositional form in a judgment, (indeed, it may operate upon any judgment whatsoever) and in conceiving this fact to consist in the relation between the subject of that judgment and another subject which has a mode of being that merely consists in the truth of propositions of which the corresponding concrete term is the predicate. Thus, we transform the proposition, 'Honey is sweet' into 'honey possesses sweetness.' (29)

Like Scotus, Peirce must face up to the question of whether his abstractions are mere artifacts of thought and language, or whether they have any objective reality. We have just seen that his realism about Firsts (what *may* be) affords him one basis for a positive answer; his realism about Thirds (what *would* be) affords him another. Recall that it is this element of his "extreme realism" that sets it apart from the "halting realism of Scotus" (p. xxxv above). For Scotus the ultimately real thing is the concrete individual (Second). Peirce is also a realist about Seconds; he regularly faults Hegel and other idealists for their neglect of Secondness (Peirce 1885, E1.233). But as Boler (1963, 138–143) points out, Peirce tends to drain Seconds of their content and restrict the reality of Secondness to instantaneous reactions; what we ordinarily count as an individual—an individual person, say—is not a Second but rather a Third, a law that governs the host of Seconds that go to make up the person (Peirce 1903f, E2.221–222). When Peirce defines an abstraction as "a substance whose being consists in the truth of some proposition concerning a more primary substance" he sounds for a moment like an orthodox scholastic realist for whom the individual, the primary substance, is truly primary; but his heterodoxy promptly slips out, when he adds that "whether there is any [truly] primary substance . . . or not we may leave the metaphysicians to wrangle about" (S9, 73).[39] Soon thereafter (75), he points out that macroscopic bodies are themselves abstractions for a believer in the atomic constitution of matter. So Peirce's more extremely realistic defense of mathematical objects is that we have precious little reality left over once we demote abstractions to the ontological second class.

Peirce's two-fold modal realism, which commits him to the reality of both Firsts and Thirds, gives rise to a corresponding complexity in his modal theory of mathematical structure. It is the truth, but not the whole Peircean truth, to say that the natural number structure is a First, whose being consists in its possibility. His metaphysical characterization of that structure in selection 15 is more categorially mixed: he calls it a "cluster of ideas of individual things" (117).[40] That is, it is a law-governed complex of possibilia; in the nomenclature he is making every effort to avoid in that text, it is a Third of Firsts. Since Thirdness is the category of law, we might say that Peirce goes Quine one better and holds that numbers *are* their laws. But the original quip reminds us that Peirce still owes us an explanation of how those laws are known. Benacerraf's Dilemma will not go quietly: we can learn about the laws that govern diamonds by scratching them, but how do we scratch the natural numbers? We have already seen Peirce's answer: the comparison between chemical and mathematical experimentation that served as our introduction to his metaphysics for mathematics turns out to speak directly to our epistemological quandary. What is it that corresponds, on the mathematical side of the comparison, to the sample through which the chemist studies the molecular structure that is the ultimate object of her inquiry? It is the diagram, which has the very "form of relation" that the mathematician wishes to study. Peirce's semiotics play a major role here: in selection 10 he stresses, as he does elsewhere, that a mathematical diagram is an icon, which represents its object by resembling it. In his writings on semiotics he strongly associates icons with the category of Firstness: icons are uniquely well-adapted to the representation of Firsts; indeed, a pure icon is itself a First.[41] There is much more to be said about Peirce's semiotic analysis of mathematical practice; I will have to leave most of that to him, and to other commentators.[42] But two signal advantages of icons, for the study of mathematical structures, are deserving of our notice here. The first is that the object of an icon need not exist; so this is just the kind of sign one needs to represent "hypothetical states of things" that may not be realized. The second is that if the icon resembles the hypothetical state, then in manipulating the icon and noticing what changes result, we can learn about the results of corresponding changes in the state of things that the icon represents.[43] It is thus the iconic nature of the diagram that entitles us to say that the mathematician, no less than the chemist, operates directly on the object of inquiry as she conducts her experiments.

But is it not highly implausible that a single diagram—the hypothesis of the Chinese Remainder Theorem, for instance—should somehow present the whole natural number structure to the mathematician's gaze? The implausibility diminishes somewhat if we think not of the hypothesis but of the whole theorem. That at least has the form of a law: it tells us that whenever we have two sequences of natural numbers satisfying certain conditions, there is a natural number that stands in a certain relation to them. This conditional is necessary: it tells us that whenever the antecedent is true, the consequent *must be* true. In his discussion of mathematical truth in selection 4 (27)

Peirce says, in effect, that all mathematical theorems are necessitated conditionals; this subsumes them under his general conception of laws as conditional necessities (p. xxxv above), and is also the standard modal-structuralist explication of truth in mathematics.[44]

All that is very well, but surely it is still implausible to think that a single diagram exhibits the whole natural number structure. Happily, Peirce never says that it does; what he says is that "the conventions of algebra . . . in conjunction with the writing of the equation" (S10, 81) create the isomorphism between the diagram and the "form of relation" that the mathematician studies. In selection 8 (61) he notes we cannot always sharply distinguish between conventions and fundamental assumptions about our subject matter. The associativity of addition, for example, is at once a basic fact about the natural number structure and a convention licensing certain diagrammatic transformations. A diagram's capacity to represent a structure, even in part, thus depends upon conventions and upon fundamental assumptions—that is, on axioms. And now suddenly the whole picture begins to make more sense. It is a commonplace, especially but not exclusively for structuralists, that axioms somehow capture mathematical structures. Peirce's philosophical system enables him to develop this familiar idea into a strikingly original account of what structures are and how we come to know them. A structure is embodied in the diagrams of the axioms that define it and the theorems we derive from them. It is a law, better still a system of laws, governing the diagrammatic experiments whereby we learn those laws.

There are plenty of hard questions to be asked about the last few pages, both as an interpretation of Peirce and as a philosophy of mathematics. What, for example, must Peirce (supposing this to be his view) say about the incompleteness phenomena, which make it doubtful that even our total practice of proving theorems can be said to capture a unique mathematical structure?[45] Introductions cannot go on forever, so even burning questions like this one will have to be left unanswered. I have not set out to produce a finished reading of Peirce's philosophy of mathematics. Having gone through the menu, I have sought in these closing paragraphs to serve the reader a philosophical appetizer. If all has gone well, it is time to bring out the first course.

Philosophy of Mathematics

1

[The Nature of Mathematics]

[Peirce 1895(?)b] Our first selection is an extended discussion of the nature of mathematics. Proceeding on the general principle that the definition of a science should be based on the function its practitioners perform within science as a whole, Peirce identifies as the "distinguishing characteristic of mathematics . . . that it is the scientific study of hypotheses which it first frames and then traces to their consequences." The mathematician is *not,* however, concerned with whether or not these hypotheses are true—that is a matter for the empirical scientist who makes use of the mathematician's results. A further point of contrast is the mathematician's minimal use of observation: he "observes nothing but the diagrams he himself constructs." This relative independence of observation sets mathematics apart, not just from empirical science, but also from logic and metaphysics, which both rely more heavily on observation than mathematics does. Mathematics is distinguished from other practices (such as poetry) that "frame hypotheses" by its exclusive concern with deducing the consequences of its hypotheses.

Much of the selection is devoted to the criticism of competing definitions, many of them due (or at least heavily indebted) to figures who deeply influenced Peirce himself: Aristotle, Kant and his own father, Benjamin Peirce. Throughout this critical discussion Peirce continues to emphasize the mathematician's indifference to the facts.

He begins with a dismissive treatment of the traditional definition of mathematics as the science of quantity. His rejection of this definition is deeply rooted in his mathematical heritage. The first of the mathematical chapters (pp. 183–193) in Murphey (1961) summarizes the developments in nineteenth century mathematics that did the most to undermine the traditional definition; these are all developments with which Peirce was intimately acquainted, in some cases through his own direct involvement. Algebra, which played a major undermining role, ran in Peirce's family. Benjamin Peirce's *Linear Associative Algebra* (Peirce 1870) is an important contribution to the field, which opens with a definition of mathematics (see note 10) that greatly influenced his son; that definition opens in turn, as it happens, with a dismissal of the traditional one, just as Peirce's does here (and elsewhere). Peirce himself, of course, is a great figure in the algebraic tradition in logic. Murphey also reviews developments in geometry that cast doubt on the identification of mathematics with the science of quantity; and here, as in other discussions of that definition, Peirce adduces projective geometry as a fatal counterexample. He did not just discard the older definition altogether, however; he takes a more irenic attitude in later writings: see especially selections 3, 13, and 14.

Peirce is equally dismissive of the suggestion, taken up from Kant by De Morgan and Hamilton, that mathematics is the science of space and time. But even here there turns out to be a grain of truth: in the classification of the sciences with which this selection concludes, Peirce assigns space and time to "the most abstract of the special sciences"; so there is after all a close affinity between the sciences of space and time, and mathematics, the most abstract science of all.

The last definition to receive extended treatment is that of Peirce's father Benjamin: "the science which draws necessary conclusions." Peirce argues that it follows from his father's definition that "mathematics must exclusively relate to the substance of hypotheses," but he rejects his father's claim that the framing of hypotheses for mathematical study is a logical and not a mathematical task.[1] He counters by denying "that everybody who reasons skilfully makes an application of logic." This sounds a major theme of Peirce's philosophy of mathematics: the independence of mathematics from logic. As he frequently does in this connection, Peirce notes here that metaphysics, by contrast with mathematics, *does* depend very closely upon logic. In discussing his father's definition, he focuses narrowly on mathematical hypotheses, whose formulation requires no logic because those hypotheses are not answerable to the facts, and therefore not open to logical criticism. But later in the selection he touches on another, deeper reason for the independence of mathematics: that "in the perspicuous and absolutely cogent reasonings of mathematics . . . appeals [to logic] are altogether unnecessary." The ensuing diatribe against publishers and teachers is more an application than an explanation of this dictum. Peirce does drop an important clue when he insists that mathematics is more, not less, abstract than logic; he has hinted at the reasons for this earlier on in the selection, with the remark that "logic rests upon observations of real facts." But this account of the independence of mathematics is incomplete at best: Peirce will do better in selection 2.

The selection ends with the first few levels of a classification of the sciences adapted from Comte. Sciences higher up in the tree are more abstract, and independent of those lower down, though they may "[borrow] data and suggestions from the discoveries" of those below. Mathematics accordingly winds up at the top. It is noteworthy that some kind of observation plays a role at every level.

§2. THE NATURE OF MATHEMATICS

Art. 2. As a general rule, the value of an exact philosophical definition of a term already in familiar use lies in its bringing out distinct conceptions of the function of objects of the kind defined. In particular, this is true of the definition of an extensive branch of science; and in order to assign the most useful boundaries for such a study, it is requisite to consider what part of the whole work of science has, from the nature of things, to be performed by those men who are to do that part of the work which unquestionably comes within the scope of that study; for it does not conduce to the clearness of a broad view of science to separate problems which have necessarily to be solved by the same men. Now a mathematician is a man whose services are called in when the physicist, or the engineer, or the underwriter, etc. finds himself confronted with an unusually complicated state of relations between facts and is in doubt whether or not this state of things necessarily involves a certain other relation between facts, or wishes to know what relation of a given kind is involved. He states the case to the mathematician. The latter is not at all responsible for the truth of those premises: that he is to accept. The first task before him is to substitute for the intricate, and often confused, mass of facts set before him, an imaginary state of things involving a comparatively orderly system of relations, which, while adhering as closely as possible or desirable to the given premises, shall be within his powers as a mathematician to deal with. This he terms his *hypothesis.* That work done, he proceeds to show that the relations explicitly affirmed in the hypothesis involve, as a part of any imaginary state of things in which they are embodied, certain other relations not explicitly stated.

Thus, the mathematician is not concerned with real truth, but only studies the substance of hypotheses. This distinguishes his science from every other. Logic and metaphysics make no special observations; but they rest upon observations which have been made by common men. Metaphysics rests upon observations of real objects, while logic rests upon observations of real facts about mental products, such as that, not merely according to some arbitrary hypothesis, but in every possible case, every proposition has a denial, that every proposition concerns some objects of common experience of the deliverer and the interpreter, that it applies to that some idea of familiar elements abstracted from the occasions of its excitation, and that it represents that an occult compulsion not within the deliverer's control unites that idea to those objects. All these are results of common observation, though they are put into scientific and uncommon groupings. But the mathematician observes nothing but the diagrams he himself constructs; and no occult compulsion governs his hypothesis except one from the depths of mind itself.

Thus, the distinguishing characteristic of mathematics is that it is the scientific study of hypotheses which it first frames and then traces to their consequences. Mathematics is either *applied* or *pure*. Applied mathematics treats of hypotheses in the forms in which they are first suggested by experience, involving more or less of features which have no bearing upon the forms of deduction of consequences from them. Pure mathematics is the result of afterthought by which these irrelevant features are eliminated.

It cannot be said that all framing of hypotheses is mathematics. For that would not distinguish between the mathematician and the poet. But the mathematician is only interested in hypotheses for the forms of inference from them. As for the poet, although much of the interest of a romance lies in tracing out consequences, yet these consequences themselves are more interesting in point of view of the resulting situations than in the way in which they are deducible. Thus, the poetical interest of a mental creation is in the creation itself, although as a part of this a mathematical interest may enter to a slight extent. Detective stories and the like have an unmistakable mathematical element. But a hypothesis, in so far as it is mathematical, is mere matter for deductive reasoning.

On the other hand, it is an error to make mathematics consist exclusively in the tracing out of necessary consequences. For the framing of the hypothesis of the two-way spread of imaginary quantity, and the hypothesis of Riemann surfaces were certainly mathematical achievements.[2]

Mathematics is, therefore, the study of the substance of hypotheses, or mental creations, with a view to the drawing of necessary conclusions.

Art. 3. Before the above analysis is definitively accepted, it ought to be compared with the principal attempts that have hitherto been made to define mathematics.

Aristotle's definition shows that its author's efforts were, in a general way, rightly directed; for it makes the characteristic of the science to lie in the peculiar quality and degree of abstractness of its objects. But in attempting to specify the character of that abstractness, Aristotle was led into error by his own general philosophy. He makes, too, the serious mistake of supposing metaphysics to be more abstract than mathematics.[3] In that he was wrong, since the former aims at the truth about the real world, which the latter disregards.

The Roman schoolmasters defined the mathematical sciences as the sciences of *quanta*. This definition would not have been admitted by a Greek geometer; because the Greeks were aware that the more fundamental branch of geometry treated of the intersections of unlimited planes. Still less does it accord with our present notion that as geometrical metrics is but a special problem in geometrical graphics, so geometrical graphics is but a special problem in geometrical topics.[4] The only defence the Romans offered of the definition was that the objects of the four mathematical sciences recognized by them, viz: arithmetic, geometry, astronomy, and music, are things possessing quantity. It does not seem to have occurred to them that the objects of grammar, logic, and rhetoric equally possess quantity, although this ought to have been obvious even to them.[5]

Subsequently, a different meaning was applied to the phrase "mathematics is the science of quantity." It is certainly possible to enlarge the conception of quantity so as to make it include tridimensional space, as imaginary quantity is two-dimensional and quaternions are four-dimensional. In such a way, this definition may be made coextensive with mathematics; but, after all, it does not throw so much light upon the position of mathematics among the sciences as that which is given in the last article.

De Morgan and Sir William Rowan Hamilton, influenced indirectly, as it would seem, by Kantianism, defined mathematics as the science of Time and Space, algebra being supposed to deal with Time as geometry does with Space.[6] Among the objections to this definition, the following seem to be each by itself conclusive.

1st, this definition makes mathematics a positive science, inquiring into matters of fact. For, even if Time and Space are of subjective origin, they are nevertheless objects of which one thing is true and another false.

The science of space is no more a branch of mathematics than is optics. That is to say, just as there are mathematical branches of optics, of which projective geometry is one, but yet optics as a whole is not mathematics, because it is in part an investigation into objective truth, so there is a mathematical branch of the science of space, but this has never been considered to include an inquiry into the true constitution and properties of space. Euclid terms statements of such properties *postulates*. Now by a postulate the early geometers understood, as a passage in Aristotle shows, notwithstanding a blunder which the Stagyrite here makes, as he often blunders about mathematics, a proposition which was open to doubt but of which no proof was to be attempted.[7] This shows that inquiry into the properties of space was considered to lie outside the province of the mathematician. In the present state of knowledge, systematic inquiry into the true properties of space is called for. It must appeal to astronomical observations on the one hand, to determine the metrical properties of space, to chemical experiment on the other hand to determine the dimensionality of space, and the question of the artiad or perissid character of space, or of its possible topical singularities (suggested by Clifford) remain as yet without any known methods for their investigation.[8] All this may be called Physical Geometry.

2nd, this definition erroneously identifies algebra with the science of Time. For it is an essential character of time that its flow takes place in one sense and not in the reverse sense; while the two directions of real quantity are as precisely alike as the two directions along a line in space. It is true that +1 squared gives itself while -1 squared gives the negative of itself. But there is another operation precisely as simple which performed upon -1 gives itself, and performed upon +1 gives the negative of itself. Besides, the idea of time essentially involves the notion of reaction between the inward and outward worlds: the future is the domain over which the Will has some power, the past is the domain of the powers which have gone to make Experience. The future and past which are essential parts of the idea of time cannot be otherwise accurately defined. Yet algebra does not treat of Will and Experience.

3rd, this definition leaves no room for some of the chief branches of mathematics, such as the doctrine of *N*-dimensional space, the theory of imaginaries, the calculus of logic, including probabilities, branches which it would be doing great violence to the natural classification of the sciences to separate from algebra and geometry.

4th, this definition is absurd, because it confines number to the domain of time, when time and space are, according to its own doctrine, *two* forms of intuition, so that their existence supposes number. Kant, himself, was too good a logician to make number a character of time. It is true that the cognition of number supposes time; but so does the cognition of Colorado silver mining. It no more follows that the science of number is a part of the science of time than that the science of Colorado silver mining is a part of the science of time.

5th, this definition would exclude *Quantity* from among the subjects of mathematical study. For *quantity,* according to the Kantian doctrine which this definition follows does not belong to intuition but is one of the four branches of the categories of the understanding, having precisely the same relation to time and space that Reality, Active Agency, and Modality have,—matters which certainly lie quite out of the province of mathematics.

It may be added that this definition is far from receiving any countenance from Kant, who makes mathematics relate to the categories of quantity and quality.[9]

In short, this definition is probably of all definitions of branches of science that have ever gained numerous adherents the very worst.

In 1870, Benjamin Peirce defined mathematics as "the science which draws necessary conclusions."[10] Since it is impossible to draw necessary conclusions except from perfect knowledge, and no knowledge of the real world can be perfect, it follows that, according to this definition mathematics must exclusively relate to the substance of hypotheses. My father seems to have regarded the work of treating the statements of fact brought to the mathematician and of creating from their suggestions a hypothesis as a basis of mathematical deduction to be the work of a logician, not of a mathematician. It cannot be denied that the two tasks, of framing hypotheses for deduction and of drawing the deductive conclusions are of widely different characters; nor that the former is similar to much of the work of the logician. But it is a mistake to suppose that everybody who reasons skillfully makes an application of logic. Logic is the science which examines signs, ascertains what is essential to being signs and describes their fundamentally different varieties, inquires into the general conditions of their truth, and states these with formal accuracy, and investigates the law of the development of thought, accurately states it and enumerates its fundamentally different modes of working.[11] In metaphysics, no skill in reasoning can avail, unless that reasoning is based upon the exact generalizations of the logician as premises; and this may truly be said to be an application of logic. But in framing mathematical hypotheses no logic is required, since it is indifferent from a mathematical point of view how far the hypothesis agrees with the observed facts. It is for the employer of the mathematician to decide that. The framing of a mathematical hypothesis does not, therefore, come within the province of the logician.

Perhaps the definition of Benjamin Peirce may be defended on the ground that the transformation of the suggestions of experience into exact mathematical hypotheses is effected by drawing necessary conclusions. The drawing of a necessary conclusion is by no means the simple act which it is commonly supposed to be; and among the acts of which it is made up there are some which would suffice, or nearly suffice, to transform the result of experience into a mathematical hypothesis. The reply is that the two parts of the mathematician's functions are markedly dissimilar and therefore require to be distinguished in the definition.

Mr. George Chrystall, in the Encyclopedia Britannica (9th Ed. Article, *Mathematics*), endeavors to define mathematics by describing the general characters essential to a mathematical hypothesis (or, in his language, a "mathematical conception").[12] This is an effort in the right direction. But the definition is not very clear. The principal feature insisted upon is that the mathematical hypothesis must be marked by a *finite number* of distinct specifications. This implies that by "specification" is meant some act, or element of an act, of which it is possible to make an infinite number; and what this is remains unexplained.

Art. 4. It may be objected to the definition of art. 2 that it places mathematics above logic as a more abstract science the different steps [of] which must precede those of logic; while on the contrary logic is requisite for the business of drawing necessary conclusions. But this I deny. An application [of] logical theory is only required by way of exception in reasoning, and not at all in mathematical deduction. In probable reasoning where the evidence is very insufficient, or almost entirely wanting, we often hear men appeal to the "burden of proof" and other supposed logical principles; and in metaphysics logical theory is the only guide. But in the perspicuous and absolutely cogent reasonings of mathematics such appeals are altogether unnecessary. Many teachers of geometry think that it is desirable that a course in logic should precede the study of the elements. But the reasoning of the elements of geometry is often bad, owing to the fact that the hypothesis is not fully stated. The matter being confused and not made interesting, the pupil's mind becomes confused; and the teacher, not knowing enough either of geometry or of psychology to know what the difficulty is or how to remedy it, turns to the apparatus of the traditional syllogistic, which, as he teaches it, will serve to throw dust in the eyes of pupil and of teacher and make them both fancy the difficulty conquered. Publishers are of opinion that teachers would not use books in which the hypotheses of geometry should be fully set forth, and in which by taking up topics, graphics, and metrics in their logical order, the reasoning should be rendered unimpeachable;[13] but it is certain that even dull pupils find no difficulty when they are taught in that way, so that only correct mathematical reasonings fully developed are offered to them. Undoubtedly, for any mind there is a point of complexity at which that mind will become confused, from inability to hold so many threads; and this point is very different for different minds. But such a difficulty is in no degree lessened by any appeal to the generalizations of logic.

Art. 5. In order to suggest the place which mathematics would seem to take in the system of the sciences, according to the above analysis, I venture to propose the following scheme of classification of all the sciences, modified from that of Auguste Comte, and like his proceeding from the more abstract to the more concrete.[14] Each science (except mathematics) rests upon fundamental principles drawn from the truths discovered by the science immediately preceding it in the list, while borrowing data and suggestions from the discoveries of those which follow it.

1. *Mathematics,* which observes only the creations of the mathematician himself. It borrow suggestions from all other sciences, from philosophy Mathematical Logic, from psychics Mathematical Economics, from physics Mathematical Optics, Metrics, etc.

2. *Philosophy,* which makes no special observations, but uses facts commonly known. In order to be exact, it must rest on mathematical principles. It divides into *Logic,* which studies the world of thought, and *Metaphysics,* which studies the world of being; and the latter must rest upon the principles of the former.

3. *The science of time* and the *science of space* are the most abstract of the special sciences. They must be based upon metaphysical *principles,* and geometry largely upon the science of time, though the one draws *data* from psychology, and the other from astronomy and chemistry.

The science of time is psychical and that of space physical; and from this point on Psychics and Physics are widely separated, and influence one another little.

2

The Regenerated Logic

[Peirce 1896] The security of mathematical reasoning, which was raised but not explained in selection 1, receives a much more expansive treatment here. That security manifests itself in the history of mathematics, in the absence of prolonged disagreement over any properly mathematical question. The explanation for that, according to Peirce, is that the objects of mathematics are creatures of the mind, which can be summoned and studied at will. The history of logic, by contrast, is replete with endlessly unsettled questions. Peirce concludes that logic must be given its proper grounding in mathematics, and metaphysics likewise in logic, and once again lays out a Comtean hierarchy of the sciences, with mathematics at the pinnacle of abstractness. The selection concludes with a rather rich comparative discussion of the observational bases, and relationships to fact, of mathematics and philosophy (which is here taken to comprise logic and metaphysics). Peirce insists that all of these sciences, including mathematics, *have* an observational basis. What sets mathematics apart is its independence of *external* observation; the abstractness of philosophy consists in its independence of *special* observation. Though mathematics is not altogether independent of observation, it is altogether independent of the facts: it makes no factual assertions, but only hypothetical ones. Even logic, by contrast, turns out to be a positive science, grounded in the external observation of facts of a very general and abstract sort.

It is a remarkable historical fact that there is a branch of science in which there has never been a prolonged dispute concerning the proper objects of that science. It is the mathematics. Mistakes in mathematics occur not infrequently, and not being detected give rise to false doctrine, which may continue a long time. Thus, a mistake in the evaluation of a definite integral by Laplace, in his *Mécanique céleste,* led to an erroneous doctrine about the motion of the moon which remained undetected for nearly half a century. But after the question had once been raised, all dispute was brought to a close within a year. So, several demonstrations in the first book of Euclid, notably that of the 16th proposition, are vitiated by the erroneous assumption that a part is necessarily less than its whole. These remained undetected until after the theory of the non-Euclidean geometry had been completely worked out; but since that time, no mathematician has defended them; nor could any competent mathematician do so, in view of Georg Cantor's, or even of Cauchy's discoveries.[1] Incessant disputations have, indeed, been kept up by a horde of undisciplined minds about quadratures, cyclotomy, the theory of parallels, rotation, attraction, etc. But the disputants are one and all men who cannot discuss any mathematical problem without betraying their want of mathematical power and their gross ignorance of mathematics at every step. Again, there have been prolonged disputes among real mathematicians concerning questions which were not mathematical or which had not been put into mathematical form. Instances of the former class are the old dispute about the measure of force, and that lately active concerning the number of constants of an elastic body; and there have been sundry such disputes about mathematical physics and probabilities.[2] Instances of the latter class are the disputes about the validity of reasonings concerning divergent series, imaginaries, and infinitesimals. But the fact remains that concerning strictly mathematical questions, and among mathematicians who could be considered at all competent, there has never been a single prolonged dispute.

It does not seem worth while to run through the history of science for the sake of the easy demonstration that there is no other extensive branch of knowledge of which the same can be said.

Nor is the reason for this immunity of mathematics far to seek. It arises from the fact that the objects which the mathematician observes and to which his conclusions relate are objects of his mind's own creation. Hence, although his proceeding is not infallible,—which is shown by the comparative frequency with which mistakes are committed and allowed,—yet it is so easy to repeat the inductions upon new instances, which can be created at pleasure, and extreme cases can so readily be found by which to test the accuracy of the processes, that when attention has once been directed to a process of reasoning suspected of being faulty, it is soon put beyond all dispute either as correct or as incorrect.

Hence, we homely thinkers believe that, considering the immense amount of disputation there has always been concerning the doctrines of logic, and especially concerning those which would otherwise be applicable to settle disputes concerning the accuracy of reasonings in metaphysics, the safest way is to appeal for our logical principles to the science of mathematics, where error can only long go unexploded on condition of its not being suspected.

This double assertion, first, that logic ought to draw upon mathematics for control of disputed principles, and second that ontological philosophy ought in like manner to draw upon logic, is a case under a general assertion which was made by Auguste Comte, namely, that the sciences may be arranged in a series with reference to the abstractness of their objects; and that each science draws regulating principles from those superior to it in abstractness, while drawing data for its inductions from the sciences inferior to it in abstractness. So far as the sciences can be arranged in such a scale, these relationships must hold good. For if anything is true of a whole genus of objects, this truth may be adopted as a principle in studying every species of that genus. While whatever is true of a species will form a datum for the discovery of the wider truth which holds of the whole genus. Substantially the following scheme of the sciences is given in the *Century Dictionary:*[3]

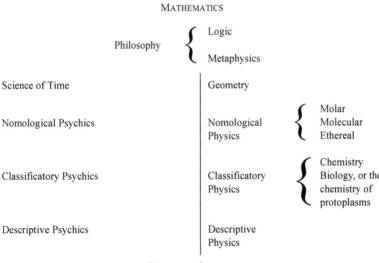

Perhaps each psychical branch ought to be placed above the corresponding physical branch. However, only the first three branches concern us here.

Mathematics is the most abstract of all the sciences. For it makes no external observations, nor asserts anything as a real fact. When the mathematician deals with facts, they become for him mere "hypotheses"; for with their truth he refuses to concern himself. The whole science of mathematics is a science of hypotheses; so that nothing could be more completely abstracted from concrete reality. Philosophy is not quite so abstract. For though it makes no *special* observations, as every other positive science does, yet it does deal with reality. It confines itself, however, to the universal phenomena of experience; and these are, generally speaking, sufficiently revealed in the ordinary observations of every-day life. I would even grant that philosophy, in the strictest sense, confines itself to such observations as *must* be open to every intelligence which can learn from experience. Here and there, however, metaphysics avails itself of one of the grander generalizations of physics, or more often of psychics, not as a governing principle, but as a mere datum for a still more sweeping generalisation. But logic is much more abstract even than metaphysics. For it does not concern itself with any facts not implied in the supposition of an unlimited applicability of language.

Mathematics is not a positive science; for the mathematician holds himself free to say that *A* is *B* or that *A* is not *B*, the only obligation upon him being, that as long as he says *A* is *B,* he is to hold to it, consistently. But logic begins to be a positive science; since there are some things in regard to which the logician is not free to suppose that they are or are not; but acknowledges a compulsion upon him to assert the one and deny the other. Thus, the logician is forced by positive observation to admit that there is such a thing as doubt, that some propositions are false, etc. But with this compulsion comes a corresponding responsibility upon him not to admit anything which he is not forced to admit.

3

The Logic of Mathematics in Relation to Education

[Peirce 1898c] The opening paragraphs of this selection cover the same ground as selection 1, with important variations. Kant receives separate treatment, and his authority is invoked in opposition to the definition of mathematics as the science of quantity, and in support of the diagrammatic nature of mathematical reasoning. The Hamilton/De Morgan definition of mathematics is now accordingly seen as a distortion of Kant's view. Kant comes up again later on, and again in connection with diagrams. Peirce insists that all necessary reasoning is diagrammatic, and faults Kant for not recognizing that this is true even of necessary reasoning in philosophy. But Kant is credited with recognizing that mathematics involves "mental experimentation" on diagrams.[1] Though Peirce does not introduce the theorematic/corollarial distinction here (but see selection 4), he does foreshadow it when he says that this experimentation reveals "new relations . . . among its parts, not stated in the precept by which it was formed."

In fact diagrams pervade this selection as a whole. They figure, for instance, in Peirce's explanation of the "necessary character" of mathematical experimentation: it "is due simply to the circumstance that the subject of this observation and experiment is a diagram of our own creation, the conditions of whose being we know all about." At first glance this seems to mistake an epistemological feature for a metaphysical one: it is plausible that we are well placed to *know* about our own mental constructions, but that does not imply the necessity of what we know about them. But bear in mind that he is talking here about constructions "the conditions of whose being we know all about." He has just concluded a detailed discussion of the abstractness of the mathematician's hypotheses; what makes them abstract is that "[a]ll features that have no bearing upon the relations of the premises to the conclusion are effaced and obliterated. The skeletonization or diagrammatization of the problem serves more purposes than one; but its principal purpose is to strip the significant relations of all disguise." The passage is somewhat obscure, but the idea appears to be that the diagram is stripped down to its essentials in such a way that any conclusions that we reach by inspecting it—or even by experimenting on it—will necessarily hold good of the hypothetical state of things that it depicts. Peirce strongly contrasts the ideality of mathematical hypotheses, which are capable of yielding necessary conclusions, with factual assertions: "in regard to the real world, we have no right to presume that any given intelligible proposition is true in absolute strictness."

Peirce takes some further steps here towards an accommodation with the definition of mathematics as the science of quantity. He admits that a "scale of quantity," a temporary quantitative scaffolding, can be of great heuristic importance for the mathematician. He also observes that there is a quantitative element even in Boole's logic; later on (e.g., in selection 9) this will be part of a quantitatively based classification of the mathematical sciences.

---

1. OF MATHEMATICS IN GENERAL

In order to understand what number is, it is necessary first to acquaint ourselves with the nature of the business of mathematics in which number is employed.

I wish I knew with certainty the precise origin of the definition of mathematics as the science of quantity. It certainly cannot be Greek, because the Greeks were advanced in projective geometry, whose problems are such as these: whether or not four points obtained in a given way lie in one plane; whether or not four planes have a point in common; whether or not two rays (or unlimited straight lines) intersect, and the like—problems which have nothing to do with quantity, as such. Aristotle names, as the subjects of mathematical study, quantity and continuity. But though he never gives a formal definition of mathematics, he makes quite clear, in more than a dozen places, his view that mathematics ought not to be defined by the things which it studies but by its peculiar mode and degree of abstractness. Precisely what he conceives this to be it would require me to go too far into the technicalities of his philosophy to explain; and I do not suppose anybody would today regard the details of his opinion as important for my purpose. Geometry, arithmetic, astronomy, and music were, in the Roman schools of the fifth century[*] and earlier, recognized as the four branches of mathematics. And we find Boëthius (A.D. 500) defining them as the arts which relate, not to quantity, but to *quantities,* or *quanta.* What this would seem to imply is, that mathematics is the foundation of the minutely exact sciences; but

---

[*]. Davidson, *Aristotle and the ancient educational ideals.* Appendix: The Seven Liberal Arts. (New York: Charles Scribner's Sons.)

really it is not worth our while, for the present purpose, to ascertain what the schoolmasters of that degenerate age conceived mathematics to be.

In modern times projective geometry was, until the middle of this century, almost forgotten, the extraordinary book of Desargues[*] having been completely lost until, in 1845, Chasles came across a MS. copy of it;[2] and, especially before imaginaries became very prominent, the definition of mathematics as the science of quantity suited well enough such mathematics as existed in the seventeenth and eighteenth centuries.

Kant, in the *Critique of pure reason* (Methodology, chapter I, section I), distinctly rejects the definition of mathematics as the science of quantity.[3] What really distinguishes mathematics, according to him, is not the subject of which it treats, but its method, which consists in studying constructions, or diagrams. That such is its method is unquestionably correct; for, even in algebra, the great purpose which the symbolism subserves is to bring a skeleton representation of the relations concerned in the problem before the mind's eye in a schematic shape, which can be studied much as a geometrical figure is studied.

But Rowan Hamilton and De Morgan, having a superficial acquaintance with Kant, were just enough influenced by the *Critique* to be led, when they found reason for rejecting the definition as the science of quantity, to conclude that mathematics was the science of pure time and pure space. Notwithstanding the profound deference which every mathematician must pay to Hamilton's opinions and my own admiration for De Morgan, I must say that it is rare to meet with a careful definition of a science so extremely objectionable as this. If Hamilton and De Morgan had attentively read what Kant himself has to say about number, in the first chapter of the *Analytic of principles* and elsewhere, they would have seen that it has no more to do with time and space than has every conception.[4] Hamilton's intention probably was, by means of this definition, to throw a slur upon the introduction of imaginaries into geometry, as a false science; but what De Morgan, who was a student of multiple algebra, and whose own formal logic is plainly mathematical, could have had in view, it is hard to comprehend, unless he wished to oppose Boole's theory of logic. Not only do mathematicians study hypotheses which, both in truth and according to the Kantian epistemology, no otherwise relate to time and space than do all hypotheses whatsoever, but we now all clearly see, since the non-Euclidean geometry has become familiar

---

[*]. Brouillon, *Proiet d'une atteinte aux événemens des rencontres du cône avec son plan,* 1639.

to us, that there *is* a real science of space and a real science of time, and that these sciences are positive and experiential—branches of physics, and so not mathematical except in the sense in which thermotics and electricity are mathematical; that is, as calling in the aid of mathematics. But the gravest objection of all to the definition is that it altogether ignores the veritable characteristics of this science, as they were pointed out by Aristotle and by Kant.

Of late decades philosophical mathematicians have come to a pretty just understanding of the nature of their own pursuit. I do not know that anybody struck the true note before Benjamin Peirce, who, in 1870,[*] declared mathematics to be "the science which draws necessary conclusions," adding that it must be defined "subjectively" and not "objectively." A view substantially in accord with his, though needlessly complicated, is given in the article Mathematics, in the ninth edition of the *Encyclopaedia Brittanica*. The author, Professor George Chrystal, holds that the essence of mathematics lies in its making pure hypotheses, and in the character of the pure hypotheses which it makes. What the mathematicians mean by a "hypothesis" is a proposition imagined to be strictly true of an ideal state of things. In this sense, it is only about hypotheses that necessary reasoning has any application; for, in regard to the real world, we have no right to presume that any given intelligible proposition is true in absolute strictness. On the other hand, probable reasoning deals with the ordinary course of experience; now, nothing like a *course of experience* exists for ideal hypotheses. Hence to say that mathematics busies itself in drawing necessary conclusions, and to say that it busies itself with hypotheses, are two statements which the logician perceives come to the same thing.

A simple way of arriving at a true conception of the mathematician's business is to consider what service it is which he is called in to render in the course of any scientific or other inquiry. Mathematics has always been more or less a trade. An engineer, or a business company (say, an insurance company), or a buyer (say, of land), or a physicist, finds it suits his purpose to ascertain what the necessary consequences of possible facts would be; but the facts are so complicated that he cannot deal with them in his usual way. He calls upon a mathematician and states the question. Now the mathematician does not conceive it to be any part of his duty to verify the facts stated. He accepts them absolutely without question. He does not in the least care whether they are correct or not. He finds, however, in almost every case that

---

[*]. In his *Linear associative algebra.*

the statement has one inconvenience, and in many cases that it has a second. The first inconvenience is that, though the statement may not at first sound very complicated, yet, when it is accurately analyzed, it is found to imply so intricate a condition of things that it far surpasses the power of the mathematician to say with exactitude what its consequences would be. At the same time, it frequently happens that the facts, as stated, are insufficient to answer the question that is put. Accordingly, the first business of the mathematician, often a most difficult task, is to frame another simpler but quite fictitious problem (supplemented, perhaps, by some supposition), which shall be within his powers, while at the same time it is sufficiently like the problem set before him to answer, well or ill, as a substitute for it.[*] This substituted problem differs also from that which was first set before the mathematician in another respect: namely, that it is highly abstract. All features that have no bearing upon the relations of the premises to the conclusion are effaced and obliterated. The skeletonization or diagrammatization of the problem serves more purposes than one; but its principal purpose is to strip the significant relations of all disguise. Only one kind of concrete clothing is permitted— namely, such as, whether from habit or from the constitution of the mind, has become so familiar that it decidedly aids in tracing the consequences of the hypothesis. Thus, the mathematician does two very different things: namely, he first frames a pure hypothesis stripped of all features which do not concern the drawing of consequences from it, and this he does without inquiring or caring whether it agrees with the actual facts or not; and, secondly, he proceeds to draw necessary consequences from that hypothesis.

Kant is entirely right in saying that, in drawing those consequences, the mathematician uses what, in geometry, is called a "construction," or in general a diagram, or visual array of characters or lines. Such a construction is formed according to a precept furnished by the hypothesis. Being formed, the construction is submitted to the scrutiny of observation, and new relations are discovered among its parts, not stated in the precept by which it was formed, and are found, by a little mental experimentation, to be such that they will always be present in such a construction. Thus, the necessary reasoning of mathematics is performed by means of observation and experiment, and its necessary character is due simply to the circumstance that the subject of this observation and experiment is a diagram of our own creation, the conditions of whose being we know all about.

---

[*]. See this well put in Thomson and Tait's *Natural philosophy,* §447.

But Kant, owing to the slight development which formal logic had received in his time, and especially owing to his total ignorance of the logic of relatives, which throws a brilliant light upon the whole of logic, fell into error in supposing that mathematical and philosophical necessary reasoning are distinguished by the circumstance that the former uses constructions. This is not true. All necessary reasoning whatsoever proceeds by constructions; and the only difference between mathematical and philosophical necessary deductions is that the latter are so excessively simple that the construction attracts no attention and is overlooked. The construction exists in the simplest syllogism in Barbara. Why do the logicians like to state a syllogism by writing the major premise on one line and the minor below it, with letters substituted for the subject and predicates? It is merely because the reasoner has to notice that relation between the parts of those premises which such a diagram brings into prominence. If the reasoner makes use of syllogistic in drawing his conclusion, he has such a diagram or construction in his mind's eye, and observes the result of eliminating the middle term. If, however, he trusts to his unaided reason, he still uses some kind of a diagram which is familiar to him personally. The true difference between the necessary logic of philosophy and mathematics is merely one of degree. It is that, in mathematics, the reasoning is frightfully intricate, while the elementary conceptions are of the last degree of familiarity; in contrast to philosophy, where the reasonings are as simple as they can be, while the elementary conceptions are abstruse and hard to get clearly apprehended. But there is another much deeper line of demarcation between the two sciences. It is that mathematics studies nothing but pure hypotheses, and is the only science which never inquires what the actual facts are; while philosophy, although it uses no microscopes or other apparatus of special observation, is really an experimental science, resting on that experience which is common to us all; so that its principal reasonings are not mathematically necessary at all, but are only necessary in the sense that all the world knows beyond all doubt those truths of experience upon which philosophy is founded. This is why the mathematician holds the reasoning of the metaphysician in supreme contempt, while he himself, when he ventures into philosophy, is apt to reason fantastically and not solidly, because he does not recognize that he is upon ground where elaborate deduction is of no more avail than it is in chemistry or biology.

I have thus set forth what I believe to be the prevalent opinion of philosophical mathematicians concerning the nature of their science. It will be found to be significant for the question of number. But were I to drop this branch of the subject without saying one word more, my criticism of the old definition, "mathematics is the science of quantity," would not be quite just. It must be admitted that quantity is useful in almost every branch of mathematics. Jevons wrote a book entitled *Pure logic, the science of quality,* which expounded, with a certain modification, the logical algebra of Boole. But it is a mistake to regard that algebra as one in which there is no system of quantity. As Boole rightly holds, there is a quadratic equation which is fundamental in it.[5] The meaning of that equation may be expressed as follows: Every proposition has one or other of two *values,* being either *true* (which gives it one value) or *false* (which gives it the other). So stated, we see that the algebra of Boole is nothing but the algebra of that system of quantities which has but two values—the simplest conceivable system of quantity. The widow of the great Boole has lately written a little book[*] in which she points out that, in solving a mathematical problem, we usually introduce some part or element into the construction which, when it has served our purpose, is removed. Of that nature is a scale of quantity, together with the apparatus by which it is transported unchanged from one part of the diagram to another, for the purpose of comparing those two parts. Something of this general description seems to be indispensable in mathematics. Take, for example, the Theorem of Pappus concerning ten rays in a plane. The demonstration of it which is now usual, that of von Staudt, introduces a third dimension; and the utility of that arises from the fact that a ray, or unlimited straight line, being the intersection of two planes, these planes show us exactly where the ray runs, while, as long as we confine ourselves to the consideration of a single plane, we have no easy method of describing precisely what the course of the ray is. Now this is not precisely a system of quantity; but it is closely analogous to such a system, and that it serves precisely the same purpose will appear when we remember that that same theorem can easily (though not *so* easily) be demonstrated by means of the barycentric calculus.[6] Although, then, it is not true that all mathematics is a science of quantity, yet it is true that all mathematics makes use of a scaffolding altogether *analogous* to a system of quantity; and quantity itself has more or less utility in every branch of mathematics which has as yet developed into any large theory.

I have only to add that the hypotheses of mathematics may be divided into those *general hypotheses* which are adhered to throughout a whole branch of mathematics, and the *particular hypotheses* which are peculiar to different special problems.

---

[*]. *The Mathematical psychology of Boole and Gratry.*

4

The Simplest Mathematics

[Peirce 1902][1] The last of our selections on the nature of mathematics is the fullest and latest; it comes from a logic text (the *Minute Logic*) and its date of 1902 makes it roughly contemporaneous with the Harvard Lectures on pragmatism and the Lowell lectures excerpted in selections 9, 12, and 13. Peirce varies and enlarges upon themes from the preceding selections: the relationship between mathematics and logic, the competing definitions of mathematics, the virtual infallibility of mathematical reasoning. The reader will be well equipped, at this point, to understand his masterful treatment of these topics here.

Two points about logic do deserve special mention. Like all of the selections so far (except for the first one) this discussion of mathematics is ancillary to a logical treatise; in particular it is preparatory, Peirce tells us at the outset, to an exposition of "certain extremely simple branches of mathematics which . . . [are of] utility in logic." Though it does not get spelled out in this excerpt, Peirce is picking up here on the remark in the previous selection that the logical "algebra of Boole is nothing but the algebra of . . . the simplest conceivable system of quantity"; this is the "simplest mathematics" that gives the selection its title. So Peirce is very serious, and very specific, here about the mathematical underpinnings of logic. At the same time—and this is the second point—he urges towards the end of the selection that the mathematical part of logic ("Formal Logic," as he calls it here) is not the only or even the most important part of the subject. What impresses him now is not the mathematical, but rather the ethical, underpinnings of logic. This is because logic is concerned with the criticism of a certain kind of conduct (namely, reasoning) and thus involves a kind of ethical evaluation. The reader may want to compare these remarks on logic and ethics with the scheme of the normative sciences which is laid out at some length in the fifth Harvard lecture on pragmatism (Peirce 1903j): there logic is the third of a trio of normative sciences—aesthetics, ethics, and logic—which are preceded in the larger hierarchy of the sciences only by mathematics and phenomenology. In the present selection he leaves aesthetics out of the picture, and singles out mathematics as the most abstract science by noting that it alone (except for ethics itself) has no need of ethics!

In addition to these variations on familiar themes, Peirce introduces two other leading ideas of his philosophy of mathematics, both of which have been discussed in the introduction: hypostatic abstraction (xxxviii) and the corollarial/theorematic distinction (xxx). Here he introduces the distinction by contrasting the mainly corollarial reasoning of the philosopher, which is concerned with words and definitions, with the mathematician's theorematic manipulation of diagrams. The contrast collapses when we notice that words function, in corollarial/philosophical reasoning, as schemata. (What Peirce says here about philosophical reasoning is worth comparing with his remarks on the same subject in selection 3.)

Peirce says that an abstraction has "a mode of being that merely consists in the truth of propositions of which the corresponding concrete term is the predicate." This sounds like nominalism, but in the introductory discussion of "dormitive virtue" he insists that there "really is in opium *something* which explains its always putting people to sleep," and in an important footnote he asserts that "even a percept is an abstraction," which makes it "difficult to maintain that all abstractions are fictions." Abstractions are, then, realities in some sense, though he does not explain very fully here in just *what* sense they are real: he has a lot more to say about this in selections 9, 12, and 13.

In this chapter, I propose to consider certain extremely simple branches of mathematics which, owing to their utility in logic, have to be treated in considerable detail, although to the mathematician they are hardly worth consideration. In Chapter IV, I shall take up those branches of mathematics upon which the interest of mathematicians is centred, but shall do no more than make a rapid examination of their logical procedure. In Chapter V, I shall treat formal logic by the aid of mathematics. There can really be little logical matter in these chapters; but they seem to me to be quite indispensable preliminaries to the study of logic.

It does not seem to me that mathematics depends in any way upon logic. It reasons, of course. But if the mathematician ever hesitates or errs in his reasoning, logic cannot come to his aid. He would be far more liable to commit similar as well as other errors there. On the contrary, I am persuaded that logic cannot possibly attain the solution of its problems without great use of mathematics. Indeed all formal logic is merely mathematics applied to logic.

SECTION 1. THE ESSENCE OF MATHEMATICS

It was Benjamin Peirce, whose son I boast myself, that in 1870 first defined Mathematics as "the science which draws necessary conclusions." This was a hard saying at the time; but today, students of the philosophy of mathematics generally acknowledge its substantial correctness.

The common definition among such people as ordinary schoolmasters still is that mathematics is the science of quantity. As this is inevitably understood in English, it seems to be a misunderstanding of a definition which may be very old,* the original meaning being that mathematics is the science of *quantities,* that is, of forms possessing quantity. We perceive that Euclid was aware that a large branch of geometry had nothing to do with measurement (unless as an aid in demonstration); and, therefore, a Greek geometer of his age (early in the third century B.C.) or later could not define mathematics as the science of that which the abstract noun quantity expresses. A line, however, was classed as *a* quantity, or *quantum,* by Aristotle and his followers;[2] so that even perspective (which deals wholly with intersections and projections, not at all with lengths) could be said to be a science of *quantities,* "quantity" being taken in the concrete sense. That this was what was originally meant by the definition 'Mathematics is the science of quantity' is sufficiently shown by the circumstance that those writers who first enunciate it, about A.D. 500, that is, Ammonius Hermiae and Boethius,[4] make astronomy and music branches of mathematics; and it is confirmed by the reasons they give for doing so.† Even Philo of Alexandria (100 B.C.), who defines mathematics as the science of ideas furnished by sensation and reflection in respect to their necessary consequences, since he includes under mathematics, besides its more essential parts, the theory of numbers and geometry, also the practical arithmetic of the Greeks, geodesy, mechanics, optics (or projective geometry), music, and astronomy, must be said to take the word mathematics in a different sense from ours.[5] That Aristotle did not regard mathematics as the science of quantity, in the modern abstract sense, is evidenced in various ways. The subjects of mathematics are, according to

---

*. From what is said by Proclus Diadochus,[3] d. A.D. 485, it would seem that the Pythagoreans understood mathematics to be the answer to the two questions 'how many' and 'how much'?

†. I regret I have not noted the passage of Ammonius to which I refer. It is probably one of the excerpts given by Brandis. My MS. note states that he gives reasons showing this to be his meaning.

him, the how much and the continuous. (See *Metaph.* K.iii.1061a33.) He referred the continuous to his category of *quantum;* and therefore he did make *quantum,* in a broad sense, the one object of mathematics.

Plato, in the Sixth book of the Republic[*], holds that the essential characteristic of mathematics lies in the peculiar kind and degree of its abstraction, greater than that of physics but less than that of what we now call philosophy; and Aristotle follows his master in this definition.[6] It has ever since been the habit of metaphysicians to extol their own reasonings and conclusions as vastly more abstract and scientific than those of mathematics. It certainly would seem that problems about God, Freedom, and Immortality[7] are more exalted than, for example, the question how many hours, minutes, and seconds would elapse before two couriers travelling under assumed conditions will come together; although I do not know that this has been proved. But that the methods of thought of the metaphysicians are, as a matter of historical fact, in any aspect, not far inferior to those of mathematics is simply an infatuation. One singular consequence of the notion which prevailed during the greater part of the history of philosophy that metaphysical reasoning ought to be similar to that of mathematics, only more so, has been that sundry mathematicians have thought themselves, as mathematicians, qualified to discuss philosophy; and no worse metaphysics than theirs is to be found.

Kant regarded mathematical propositions as synthetical judgments *a priori;* wherein there is this much truth, that they are not, for the most part, what he called analytical judgments; that is, the predicate is not, in the sense he intended, contained in the definition of the subject. But if the propositions of arithmetic, for example, are true cognitions, or even forms of cognition, this circumstance is quite aside from their mathematical truth. For all modern mathematicians agree with Plato and Aristotle that mathematics deals exclusively with hypothetical states of things, and asserts no matter of fact whatever; and further, that it is thus alone that the necessity of its conclusions is to be explained.[†] This is the true essence of mathematics; and my father's definition is in so far correct that it is impossible to reason necessarily concerning anything else than a pure hypothesis. Of course, I do not mean that if such pure hypothesis happened to be true of an actual state of things, the reasoning would thereby cease to be necessary. Only, it never could be known apodictically to be true of an actual state of things. Suppose

---

[*]. 510C to the end; but in the *Laws* his notion is improved.

[†]. A view which J. S. Mill (*Logic* II v.§§1,2) rather comically calls "the important doctrine of Dugald Stewart."

a state of things of a perfectly definite, general description. That is, there must be no room for doubt as to whether anything, itself determinate, would or would not come under that description. And suppose, further, that this description refers to nothing occult,—nothing that cannot be summoned up fully into the imagination. Assume, then, a range of possibilities equally definite and equally subject to the imagination; so that, so far as the given description of the supposed state of things is general, the different ways in which it might be made determinate could never introduce doubtful or occult features. The assumption, for example, must not refer to any matter of fact. For questions of fact are not within the purview of the imagination. Nor must it be such that, for example, it could lead us to ask whether the vowel *OO* can be imagined to be sounded on as high a pitch as the vowel *EE*. Perhaps it would have to be restricted to pure spatial, temporal, and logical relations. Be that as it may, the question whether, in such a state of things, a certain other similarly definite state of things, equally a matter of the imagination, could or could not, in the assumed range of possibility, ever occur, would be one in response to which one of the two answers *Yes* and *No* would be true, but never both. But all pertinent facts would be within the beck and call of the imagination; and consequently nothing but the operation of thought would be necessary to render the true answer. Nor, supposing the answer to cover the whole range of possibility assumed, could this be rendered otherwise than by reasoning that would be apodictic, general and exact. No knowledge of what actually is, no *positive* knowledge, as we say, could result. On the other hand, to assert that any source of information that is restricted to actual facts could afford us a necessary knowledge, that is, knowledge relating to a whole general range of possibility, would be a flat contradiction in terms.

Mathematics is the study of what is true of hypothetical states of things. That is its essence and definition. Everything in it, therefore, beyond the first precepts for the construction of the hypotheses, has to be of the nature of apodictic inference. No doubt, we may reason imperfectly and jump at a conclusion: still, the conclusion so guessed at is, after all, that in a certain supposed state of things something would necessarily be true. Conversely, too, every apodictic inference is, strictly speaking, mathematics. But mathematics, as a serious science, has, over and above its essential character of being hypothetical, an accidental characteristic peculiarity,—a *proprium,* as the Aristotelians used to say,—which is of the greatest logical interest. Namely, while all the "philosophers" follow Aristotle in holding no demonstration to be thoroughly satisfactory except what they call a "direct" demonstration, or a "demonstration why,"—by which they mean a demonstration which employs only general concepts and concludes nothing but what would be an item of a definition if all its terms were distinctly defined themselves,[8] the mathematicians, on the contrary, entertain a contempt for that style of reasoning, and glory in what the philosophers stigmatize as "mere" indirect demonstrations, or "demonstrations that." Those propositions which can be deduced from others by reasoning of the kind that the philosophers extol are set down by mathematicians as "corollaries". That is to say, they are like those geometrical truths which Euclid did not deem worthy of particular mention, and which his editors inserted with a garland, or *corolla,* against each in the margin, implying perhaps that it was to them that such honor as might attach to these insignificant remarks was due. In the theorems, or at least in all the major theorems, a different kind of reasoning is demanded. Here, it will not do to confine oneself to general terms. It is necessary to set down, or to imagine, some individual and definite schema, or diagram;—in geometry, a figure composed of lines with letters attached; in algebra an array of letters of which some are repeated. This schema is constructed so as to conform to a hypothesis set forth in general terms in the thesis of the theorem. Pains are taken so to construct it that there would be something closely similar in every possible state of things to which the hypothetical description in the thesis would be applicable, and furthermore to construct it so that it shall have no other characters which could influence the reasoning. How it can be that, although the reasoning is based upon the study of an individual schema, it is nevertheless necessary, that is, applicable to all possible cases, is one of the questions we shall have to consider. Just now, I wish to point out that after the schema has been constructed according to the precept virtually contained in the thesis, the assertion of the theorem is not evidently true, even for the individual schema; nor will any amount of hard thinking of the philosophers' corollarial kind ever render it evident. Thinking in general terms is not enough. It is necessary that something should be DONE. In geometry, subsidiary lines are drawn. In algebra permissible transformations are made. Thereupon, the faculty of observation is called into play. Some relation between the parts of the schema is remarked. But would this relation subsist in every possible case? Mere corollarial reasoning will sometimes assure us of this. But, generally speaking, it may be necessary to draw distinct schemata to represent alternative possibilities. Theorematic reasoning invariably depends upon experimentation with individual schemata. We shall find that, in the last analysis, the same thing is true of the corollarial reasoning, too; even the Aristotelian "demonstration why." Only, in this case, the very words serve as schemata. Accordingly, we may say that corollarial, or "philosophical," reasoning is reasoning with words; while theorematic, or mathematical reasoning proper, is reasoning with specially constructed schemata.

Another characteristic of mathematical thought is the extraordinary use it makes of abstractions. Abstractions have been a favorite butt of ridicule in modern times. Now it is very easy to laugh at the old physician who is represented as answering the question, why opium puts people to sleep, by saying that it is because it has a dormitive virtue.[9] It is an answer that no doubt carries vagueness to its last extreme. Yet, invented as the story was to show how little meaning there might be in an abstraction, nevertheless the physician's answer does contain a truth that modern philosophy has generally denied: it does assert that there really is in opium *something* which explains its always putting people to sleep. This has, I say, been denied by modern philosophers generally. Not, of course, explicitly; but when they say that the different events of people going to sleep after taking opium have really nothing in common, but only that the mind classes them together;—and this is what they virtually do say in denying the reality of generals;—they do implicitly deny that there is any true explanation of opium's generally putting people to sleep.

Look through the modern logical treatises, and you will find that they almost all fall into one or other of two errors, as I hold them to be, that of setting aside the doctrine of abstractions, in the sense in which an abstract noun marks an abstraction, as a grammatical topic with which the logician need not particularly concern himself, and that of confounding abstraction, in this sense, with that operation of the mind by which we pay attention to one feature of a percept to the disregard of others. The two things are entirely disconnected. The most ordinary fact of perception, such as 'it is light' involves *precisive* abstraction, or *prescission*. But *hypostatic* abstraction, the abstraction which transforms 'it is light' into 'there is light here,' which is the sense which I shall commonly attach to the word abstraction (since *prescission* will do for precisive abstraction,) is a very special mode of thought. It consists in taking a feature of a percept or percepts, after it has already been prescinded from the other elements of the percept, so as to take propositional form in a judgment (indeed, it may operate upon any judgment whatsoever), and in conceiving this fact to consist in the relation between the subject of that judgment and another subject which has a mode of being that merely consists in the truth of propositions of which the corresponding concrete term is the predicate. Thus, we transform the proposition, 'Honey is sweet' into 'honey possesses sweetness.' 'Sweetness' might be called a fictitious thing, in one sense. But since the mode of being attributed to it *consists* in no more than the fact that some things are sweet, and it is not pretended, or even imagined, that it has any other mode of being, there is, after all, no fiction. The only profession made is that we can consider the fact of honey being sweet under the form of a relation; and so we really can. I have selected *sweetness* as an instance of one of the least useful of abstractions. Yet even this is convenient. It facilitates such thoughts as that the sweetness of honey is particularly cloying; that the sweetness of honey is something like the sweetness of a honeymoon; etc. Abstractions are particularly congenial to mathematics. Every day life first, for example, found the need of that class of abstractions which we call *collections.* Instead of saying that some human beings are males and all the rest females, it was found convenient to say that *mankind* consists of the male *part* and the female *part.* The same thought makes classes of collections, such as pairs, leashes, quatrains, hands, weeks, dozens, baker's dozens, sonnets, scores, quires, hundreds, long hundreds, gross, reams, thousands, myriads, lacs, millions, milliards, milliasses, etc. These have suggested a great branch of mathematics.[*] Again, a point moves: it is by abstraction that the geometer says that it "describes a line." This line, though an abstraction, itself moves; and this is regarded as generating a surface; and so on. So likewise, when the analyst treats operations as themselves subjects of operations, a method whose utility will not be denied, this is another instance of abstraction. Maxwell's notion of a tension exercized upon lines of electrical force, transverse to them, is somewhat similar.[10] These examples exhibit the great rolling billows of abstraction in the ocean of mathematical thought; but when we come to a minute examination of it, we shall find, in every department, incessant ripples of the same form of thought, of which the examples I have mentioned give no hint.

Another characteristic of mathematical thought is that it can have no success where it cannot generalize. One cannot, for example, deny that chess is mathematics, after a fashion; but, owing to the exceptions which everywhere confront the mathematician in this field, such as the limits of the board; the single steps of king, knight, and pawn; the finite number of squares; the peculiar mode of capture by pawns; the queening of pawns; castling; there results a mathematics whose wings are effectually clipped, and

---

[*]. Of course, the moment a collection is recognized as an abstraction we have to admit that even a percept is an abstraction or represents an abstraction, if matter has parts. It therefore becomes difficult to maintain that all abstractions are fictions.

which can only run along the ground. Hence it is that a mathematician often finds what a chess-player might call a gambit to his advantage; exchanging a smaller problem that involves exceptions for a larger one free from them. Thus, rather than suppose that parallel lines, unlike all other pairs of straight lines in a plane, never meet, he supposes that they intersect at infinity. Rather than suppose that some equations have roots while others have not, he supplements real quantity by the infinitely greater realm of imaginary quantity. He tells us with ease how many inflexions a plane curve of any description has; but if we ask how many of these are real, and how many merely fictional, he is unable to say. He is perplexed by three-dimensional space, because not all pairs of straight lines intersect; and finds it to his advantage to use quaternions which represent a sort of four-fold continuum, in order to avoid the exception.[11] It is because exceptions so hamper the mathematician that almost all the relations with which he chooses to deal are of the nature of correspondences; that is to say, such relations that for every relate there is the same number of correlates, and for every correlate the same number of relates.

Among the minor, yet striking characteristics of mathematics, may be mentioned the fleshless and skeletal build of its propositions; the peculiar difficulty, complication, and stress of its reasonings; the perfect exactitude of its results; their broad universality; their practical infallibility. It is easy to speak with precision upon a general theme. Only, one must commonly surrender all ambition to be certain. It is equally easy to be certain. One has only to be sufficiently vague. It is not so difficult to be pretty precise and fairly certain at once about a very narrow subject. But to reunite, like mathematics, perfect exactitude, and practical infallibility, with unrestricted universality is remarkable. But it is not hard to see that all these characters of mathematics are inevitable consequences of its being the study of hypothetical truth.

It is difficult to decide between the two definitions of mathematics, the one by its method, that of drawing necessary conclusions, the other by its aim and subject matter, as the study of hypothetical states of things. The former makes, or seems to make, the deduction of the consequences of hypotheses the sole business of the mathematician, as such. But it cannot be denied that immense genius has been exercized in the mere framing of such general hypotheses as the field of imaginary quantity and the allied idea of Riemann's surface, in imagining non-Euclidean measurement, ideal numbers, the perfect liquid. Even the framing of the particular hypotheses of special problems almost always calls for good judgment and knowledge, and sometimes for great intellectual power, as in the case of Boole's logical algebra. Shall we exclude this work from the domain of mathematics? Perhaps the answer should be that, in the first place, whatever exercise of intellect may be called for in applying mathematics to a question not propounded in mathematical form is certainly not pure mathematical thought; and in the second place, that the mere creation of a hypothesis may be a grand work of poietic genius, but cannot be said to be scientific, inasmuch as that which it produces is neither true nor false, and therefore is not knowledge. This reply suggests the further remark that if mathematics is the study of purely imaginary states of things, poets must be great mathematicians, especially that class of poets who write novels of intricate and enigmatical plots. Even the reply, which is obvious, that by *studying* imaginary states of things we mean studying what is true of them, perhaps does not fully meet the objection. The article *Mathematics* in the ninth edition of the Encyclopaedia Britannica makes mathematics consist in the study of a particular sort of hypotheses, namely, those that are exact, etc., as there set forth at some length. The article is well worthy of consideration.

The philosophical mathematician, Dr. Richard Dedekind, holds mathematics to be a branch of logic.[12] This would not result from my father's definition, which runs, not that mathematics is the science *of drawing* necessary conclusions,—which would be deductive logic,—but that it is the science *which draws* necessary conclusions. It is evident, and I know as a fact, that he had this distinction in view. At the time when he thought out this definition, he, a mathematician, and I, a logician, held daily discussions about a large subject which interested us both; and he was struck, as I was, with the contrary nature of his interest and mine in the same propositions. The logician does not care particularly about this or that hypothesis or its consequences, except so far as these things may throw a light upon the nature of reasoning. The mathematician is intensely interested in efficient methods of reasoning, with a view to their possible extension to new problems; but he does not, *quâ* mathematician, trouble himself minutely to dissect those parts of this method whose correctness is a matter of course. The different aspects which the algebra of logic will assume for the two men is instructive in this respect. The mathematician asks what value this algebra has as a calculus. Can it be applied to unravelling a complicated question? Will it, at one stroke, produce a remote consequence? The logician does not wish the algebra to have that character. On the contrary, the greater number of distinct logical steps into which the algebra breaks up an inference will for him constitute a superiority of it over another which moves more swiftly to its conclusions. He demands that the algebra shall analyze a reasoning into its last elementary steps. Thus, that which is a merit in a logical algebra for one of these students is a demerit in the eyes of the other. The one studies the science of drawing conclusions, the other the science which draws necessary conclusions.

But, indeed, the difference between the two sciences is far more than that between two points of view. Mathematics is purely hypothetical: it produces nothing but conditional propositions. Logic, on the contrary, is categorical in its assertions. True, it is not merely, or even mainly, a mere discovery of what really is, like metaphysics. It is a normative science. It thus has a strongly mathematical character, at least in its methodeutic division;[13] for here it analyzes the problem of how, with given means, a required end is to be pursued. This is, at most, to say that it has to call in the aid of mathematics, that it has a mathematical branch. But so much may be said of every science. There is a mathematical logic, just as there is a mathematical optics and a mathematical economics. Mathematical logic is formal logic. Formal logic, however developed, is mathematics. Formal logic, however, is by no means the whole of logic, or even its principal part. It is hardly to be reckoned as a part of logic proper. Logic has to define its aim; and in doing so is even more dependent upon ethics, or the philosophy of aims, by far, than it is, in the methodeutic branch, upon mathematics. We shall soon come to understand how a student of ethics might well be tempted to make his science a branch of logic; as, indeed, it pretty nearly was in the mind of Socrates. But this would be no truer a view than the other. Logic depends upon mathematics; still more intimately upon ethics; but its proper concern is with truths beyond the purview of either.[14]

There are two characters of mathematics which have not yet been mentioned, because they are not exclusive characteristics of it. One of these, which need not detain us, is that mathematics is distinguished from all other sciences except only ethics, in standing in no need of ethics. Every other science, even logic,—logic, especially,—is in its early stages in danger of evaporating in airy nothingness, degenerating, as the Germans say, into an arachnoid film, spun from the stuff that dreams are made of.[15] There is no such danger for pure mathematics; for that is precisely what mathematics ought to be.

The other character,—and of particular interest it is to us just now,—is that mathematics, along with ethics and logic alone of the sciences, has no need of any appeal to logic. No doubt, some reader may exclaim in dissent to this, on first hearing it said. Mathematics, they may say, is preeminently a science of reasoning. So it is; preeminently a science that reasons. But just as it is not necessary, in order to talk, to understand the theory of the formation of vowel sounds, so it is not necessary, in order to reason, to be in possession of the theory of reasoning. Otherwise, plainly, the science of logic could never be developed. The contrary objection would have more excuse, that no science stands in need of logic, since our natural power of reason is enough. Make of logic what the majority of treatises in the past have made of it, and a very common class of English and French books still make of it, that is to say, mainly formal logic, and that formal logic represented as an art of reasoning, and in my opinion this objection is more than sound; for such logic is a great hindrance to right reasoning. It would, however, be aside from our present purpose to examine this objection minutely; I will content myself with saying that undoubtedly our natural power of reasoning is enough, in the same sense that it is enough, in order to obtain a wireless transatlantic telegraph, that men should be born. That is to say, it is bound to come sooner or later. But that does not make research into the nature of electricity needless for gaining such a telegraph. So likewise if the study of electricity had been pursued resolutely, even if no special attention had ever been paid to mathematics, the requisite mathematical ideas would surely have been evolved. Faraday, indeed, did evolve them without any acquaintance with mathematics. Still, it would be far more economical to postpone electrical researches, to study mathematics by itself, and then to apply it to electricity, which was Maxwell's way.[16] In this same manner, the various logical difficulties which arise in the course of every science except mathematics, ethics, and logic, will, no doubt, get worked out after a time, even though no special study of logic be made. But it would be far more economical to make first a systematic study of logic. If anybody should ask what are these logical difficulties which arise in all the sciences, he must have read the history of science very irreflectively. What was the famous controversy concerning the measure of force but a logical difficulty? What was the controversy between the uniformitarians and the catastrophists but a question of whether or not a given conclusion followed from acknowledged premisses?[17] This will fully appear in the course of our studies in the present work. But it may be asked whether mathematics, ethics, and logic have not encountered similar difficulties. Are the doctrines of logic at all settled? Is the history of ethics anything but a history of controversy? Have no logical errors been committed by mathematicians? To that I reply, first, as to logic, that not only have the rank and file of writers on the subject been, as an eminent psychiatrist, Maudsley, declares, men of arrested brain-development,[18] and not only have they generally lacked the most essential qualification for the study, namely mathematical training, but the main reason why logic is unsettled is that thirteen different opinions are current as to the true aim of the science. Now this is not a logical difficulty, but an ethical difficulty; for ethics is the science of aims. Secondly, it is true that pure ethics has been, and always must be, a theatre of discussion for the reason that its study consists in the gradual development of a distinct recognition of a satisfactory aim. It is a science of subtleties, no doubt; but it is not logic, but the development of the ideal, which really creates and resolves the problems of ethics. Thirdly, in mathematics errors of reasoning have occurred, nay, have passed unchallenged for thousands of years. This, however, was simply because they escaped notice. Never, in the whole history of the science, has a question whether a given conclusion followed *mathematically* from given premises, once started, failed to receive a speedy and unanimous reply. Very few have been even the apparent exceptions; and those few have been due to the fact that it is only within the last half century that mathematicians have come to have a perfectly clear recognition of what is mathematical soil and what foreign to mathematics. Perhaps the nearest approximation to an exception was the dispute about the use of divergent series.[19] Here neither party was in possession of sufficient pure mathematical reasons covering the whole ground; and such reasons as they had were not only of an extra-mathematical kind, but were used to support more or less vague positions. It appeared then, as we all know now, that divergent series are of the utmost utility.[20] Struck by this circumstance, and making an inference of which it is sufficient to say that it was not mathematical, many of the old mathematicians pushed the use of divergent series beyond all reason. This was a case of mathematicians disputing about the validity of a kind of inference that is not mathematical. No doubt, a sound logic (such as has not hitherto been developed) would have shown clearly that that non-mathematical inference was not a sound one. But this is, I believe, the only instance in which any large party in the mathematical world ever proposed to rely, in mathematics, upon unmathematical reasoning. My proposition is that true mathematical reasoning is so much more evident than it is possible to render any doctrine of logic proper,—without just such reasoning,—that an appeal in mathematics to logic could only embroil a situation. On the contrary, such difficulties as may arise concerning necessary reasoning have to be solved by the logician by reducing them to questions of mathematics. Upon those mathematical dicta, as we shall come clearly to see, the logician has ultimately to repose.

So a double motive induces me to devote some preliminary chapters to mathematics. For, in the first place, in studying the theory of reasoning, we are concerned to acquaint ourselves with the methods of that prior science of which acts of reasoning form the staple. In the second place, logic, like any other science, has its mathematical department, and of that, a large portion, at any rate, may with entire convenience be studied as soon as we take up the study of logic, without any propedeutic. That portion is what goes by name of Formal Logic.[*] It so happens that the special kind of mathematics needed for formal logic, which, therefore, we need to study in detail, as we need not study other branches of mathematics, is so excessively simple as neither to have much mathematical interest, nor to display the peculiarities of mathematical reasoning. I shall, therefore, devote the present chapter,—a very dull one, I am sorry to say, it must be,—to this kind of mathematics. Chapter IV will treat of the more truly mathematical mathematics; and Chapter V will apply the results of the present chapter to the study of Formal Logic.

---

[*]. "Formal Logic" is also used, by Germans chiefly, to mean that sect of logic which makes Formal Logic pretty much the whole of Logic.

5

The Essence of Reasoning

[Peirce 1893a] The ideality of mathematics is the center of attention in this brief but very rich excerpt from the *Grand Logic*. Peirce compares arithmetic, a branch of pure mathematics, with the applied fields of logic and geometry. Logic is "intermediate" between the other two: unlike arithmetic, it is concerned with "questions of fact," but unlike geometry, it "knows nothing about the truths of nature." Arithmetic and geometry are more sharply separated. Space, the subject matter of geometry, is "a matter of real experience," and geometrical concepts like straightness and length involve vision and "the sense of muscular action." Moreover, some geometrical assertions at least are conceivably open to experimental refutation. With arithmetic, on the other hand, this is quite inconceivable. An apparent counterexample to some basic number theoretic truths is seen to fail because it does not "conform to the idea of number." Here Peirce takes his stand with Frege, over against Mill and the empiricist tradition, in denying that the truths of arithmetic are empirical generalizations about the behavior of physical objects (Frege 1884, 12–13).

The truths of mathematics are truths about ideas merely. They are all but certain. Only blundering can introduce error into mathematics. Questions of logic are questions of fact. Can the premise be true and the conclusion be false at the same time? But the logician, as such, knows nothing about the truth of nature. He only hopes that a few assumptions he makes may be near enough correct to answer his purpose in some measure. These assumptions are, for instance, that things are sufficiently steady for something to be true, and what contradicts it false, that nothing is true and false at once, etc.

The assertion that mathematics is purely ideal requires some explanation. Thomson and Tait (Natural Philosophy §438) wisely remark that it is "utterly impossible to submit to mathematical reasoning the exact conditions of any physical question."[1] A practical problem arises, and the physicist endeavors to find a soluble mathematical problem that resembles the practical one as closely as it may. This involves a logical analysis of the problem, a putting of it into equations. The mathematics begins when the equations, or other purely ideal conditions, are given. "Applied mathematics" is simply the study of an idea which has been constructed so as to be more or less like nature. Geometry is an example of such applied mathematics; although the mathematician often makes use of space imagination to form icons of relations which have no particular connection with space. This is done, for example, in the theory of equations and throughout the theory of functions. And besides such special applications of geometrical ideas, all a mathematician's diagrams are visually imagined, and involve space. But space is a matter of real experience; and when it is said that a straight line is the shortest distance between two points, this cannot be resolved into a merely formal phrase, like 2 and 3 are 5. A straight line is a line that viewed endwise appears as a point; while length involves the sense of muscular action. Thus the connection of two experiences is asserted in the proposition that the straight line is the shortest. But 2 and 3 are 5 is true of an idea only, and of real things so far as that idea is applicable to them. It is nothing but a form, and asserts no relation between outward experiences. If to a candle, a book, and a shadow,—three objects, is joined a book, one, the result is 5, because there will be two shadows; and if 5 more candles be brought, the total will be only 8, because the shadows are destroyed. But nobody would take such facts as violations of arithmetic; for the propositions of arithmetic are not understood as applicable to matters of fact, except so far as the facts happen to conform to the idea of number. But it is quite possible that if we could measure the angles of a triangle with sufficient accuracy, we should find they did not sum up to 180°, but either exceeded it or fell short. There is no difficulty in conceiving this; although, owing to numerous associations of ideas, it is necessary to devote some weeks to a careful study of the matter before it becomes perfectly clear. Accordingly, geometrical propositions and arithmetical propositions stand upon altogether different ground. The footing of logical principles is intermediate between these. The logician does not assert anything, as the geometrician does; but there are certain assumed truths which he hopes for, relies upon, banks upon, in a way quite foreign to the arithmetician. Logic teaches us to expect some residue of dreaminess in the world, and even self-contradictions; but we do not expect to be brought face to face with any such phenomenon, and at any rate are forced to run the risk of it. The assumptions of logic differ from those of geometry, not merely in not being assertorically held, but also in being much less definite.

6

New Elements of Geometry

[Peirce 1894] Like the previous selection, this is an excerpt from a textbook: Peirce's revision of his father's geometry text, which was rejected by Ginn and Company because "the mathematical philosophy [in it] would meet with hardly any sale" (Brent 1993, 242). Peirce begins by observing that the mathematician's diagrams have a tendency to take on a life of their own: to "generalize" beyond their original intended interpretations. Though he does not mention the fact here, his own algebraic tradition in logic is a prime example (Murphey 1961, 183–185). The selection concludes with a tantalizing discussion of the objectivity of mathematics, and of mathematical surprise, by way of an analogy between mathematics and the "occult science" of chemistry. That analogy is developed more thoroughly in selection 10; for more on the mathematician's discovery of "occult qualities," not in external things but in her own ideas, see selections 7 and 14, and compare the various contexts of the word 'occult' on pp. xxvii, 3, and 27.

We see that in certain cases, namely when $C > B^2$, the expressions for the roots of the quadratic

$$X_1 = -B + \sqrt{B^2 - C}$$

$$X_2 = -B - \sqrt{B^2 - C}$$

do not represent any calculable numbers.

Nevertheless, according to the rules of algebra, these expressions satisfy the equation. Thus

$$X_1^2 = 2B^2 - 2B\sqrt{B^2 - C} - C.$$

$$X_2^2 = 2B^2 + 2B\sqrt{B^2 - C} - C.$$

$$2BX_1 = -2B^2 + 2B\sqrt{B^2 - C}.$$

$$2BX_2 = -2B^2 - 2B\sqrt{B^2 - C}.$$

So that

$$X_1{}^2 + 2BX_1 + C = 0$$

$$X_2{}^2 + 2BX_2 + C = 0$$

As "literal" (i.e., formed of letters) expressions, they present nothing extraordinary. But there are no corresponding numbers.

Here we have an instance of a phenomenon frequent in mathematics, namely, that our diagrams, or our formulae, or whatever it may be we are thinking with, insist upon being generalized.[*] An opinion came to be prevalent not very long ago that mathematical thought was entirely of the nature of deducing consequences from express principles, like this:

All men must die,

And there[fore] I must die.

This opinion involves a number of mistakes about reasoning, in general; and it also involves a most important error about mathematics. It is true that exact proof, the tracing out of inevitable consequences, not usually of "principles" exactly, but of arrangements, plays a great part in mathematics,—so much more than in any other sciences, that it stamps mathematics with a peculiar character. Nevertheless, that which makes the gist of mathematical thought, and which constitutes its great interest, is something in which it resembles other sciences, while resembling them with a difference.

If the student has gone over the preceding pages of this book, he cannot but recognize, when it is pointed out to him, that the very life of mathematical thinking consists in making experiments upon diagrams and the like and in observing the results.

The chemist mixes two colorless liquids and finds the mixture a brilliant blue or red. Such discoveries caused the middle ages to term chemistry an "occult" science; that is, it brought to light results which no reasoning could anticipate. In reasoning like "All men must die, therefore so must I," the con-

---

*. *Diagram* is a word which will do for any visual skeleton form in which the relations of parts are perspicuously exhibited, and are distinguished by lettering or otherwise, and which has some signification, or at least some significance. A system of equations written under one another so that their relations may be seen at a glance may well be called a diagram. Indeed any algebraical expression is essentially a diagram.

clusion contains no idea not plainly existing in the premises. The conclusion is formed of the thoughts in the premises.

Now mathematical phenomena are just as unexpected as any results of chemical experiments. They do not startle the senses, but they involve new ideas.

Although mathematics deals with ideas and not with the world of sensible experience, its discoveries are not arbitrary dreams but something to which our minds are forced and which were unforeseen.

Besides experiment, the natural sciences make the greatest use of *generalization.* In mathematics it plays fully as prominent a part; all the great steps depend upon it; and it leads to even more wonderful enlargements of our conceptions.

These remarks cannot be fully appreciated by the student at this time; but they will be useful, in directing his attention to the manner in which ideas spring up in mathematics, and grow, and put forth flowers. To watch this process is the most interesting part of the study of algebra and of geometry.

7

On the Logic of Quantity

[Peirce 1895(?)c] The manuscript from which this selection is taken contains fragments of more than one version of the same treatise; hence the multiple treatments of diagrams, and of the place of probable reasoning in mathematics. The discussion of diagrams contains, along with material by now familiar, a strong affirmation of their visual character. At the same time Peirce insists that a single hypothesis or system of relations can be embodied in multiple diagrams; he will pursue this idea further in selection 10. Another important refinement is the distinction between inherential and imputational diagrams. This blocks an obvious objection to his sweeping generalization that all mathematical reasoning is diagrammatic; for even if we concede that geometrical reasoning relies on diagrams, this is less obviously true for algebraic reasoning. In the geometrical case the system of relations is directly seen—it is inherent—in the relation of the parts of the diagram. In the algebraic case the relations are imputed to the diagram by way of conventions and rules: for example, the commutativity of a group is represented by the rule that licenses transposition of terms under the appropriate conditions. The distinction between the two kinds of diagrams is briefly stated, and appears to be fairly hard and fast, but in other selections (especially 8 and 10) Peirce comes around to a more nuanced view, and accords a wider range of operation to convention.

The main topic of the projected memoir is quantity; though Peirce does not directly mention the traditional definition of mathematics as the science of quantity, he once again attempts to articulate what is right about it. Each branch of mathematics is constituted by what he calls a "general hypothesis," which specifies a system of objects with certain relations to one another, for example, the real numbers under the usual ordering and arithmetical relations, or the points of a space satisfying the Euclidean axioms. The special hypothesis of a particular problem within a given branch concerns a system of relations specified in terms of the objects and relations of the background structure. As Peirce puts it, the background structure laid out by the general hypothesis serves as a "middle term of comparison" for the system of relations laid down in the special hypothesis of the particular problem. Any background structure that serves as a "middle term" in this way

Peirce calls a *system of quantity;* we have already encountered this analysis of quantity in selection 3, which was written three years later than this one.

In both versions of his introductory remarks Peirce takes up the question of probable reasoning in mathematics. Strictly speaking there can be no such thing because such reasoning seeks to discover the limiting value that a certain proportion (e.g., the proportion of coin flips that come up heads) approaches over the course of our experience. But the objects of mathematics—hypothetical systems of relations—are not experienced in the appropriate way and hence are not subject to probable reasoning. Our own processes of mathematical discovery, by contrast, *are* experienced and hence can be reasoned about probabilistically; Peirce gives the example of the computation, and re-checking, of a complex sum. In one version of his treatise but not the other Peirce suggests that conjectures about an infinite set, based on a finite sample, might also count as probable mathematical reasoning; yet he all but dismisses such reasoning as "excessively weak" because there is no such thing as true random sampling in mathematics, where any (finite) sample is dwarfed by the infinite population under study.

The bulk of the selection is given over to a brilliant epistemological analysis of mathematics, informed by Peirce's semiotics and his triadic system of categories. This exposition of the categories is atypical and, in one important respect, incomplete. The initial approach is unusually mentalistic, and Peirce has to contort, and in places to distort, his categories to fit them into an associationist framework. When allowances are made for that, the categories of Firstness and Secondness come through clearly enough, but Thirdness is reduced to one of its effects, namely, the clustering of ideas. It comes through a bit more clearly at the end, when Peirce strips away the mentalistic scaffolding.

The application to mathematics begins with the observation that "clustering of ideas is either due to an outward occult power or to an inward one." Peirce recognizes two such occult powers, Nature and Reason, which respectively cause the clustering of our ideas of the outer and inner worlds. It is just here that one would have expected Thirdness to be brought front and center, but except for a passing reference to "mediation," Peirce continues to keep it under wraps. It is a defining tenet of his realism that there is genuine Thirdness in Nature, so it is natural to suppose that his thought here is that there is genuine Thirdness in Reason as well; this would be to go beyond, though not very far beyond, the letter of the text. Our capacity to be surprised by mathematical discoveries is offered as evidence for the "occult" nature of Reason. The properties of mathematical structures are discovered rather than constituted by mathematical reasoning. On this interpretation, Peirce holds to a realism about mathematics that is simply the translation, to the inner world, of a major component of his realism about the outer world.

The strongest justification for the translation would of course be a significant overlap between the Thirdnesses of the inner and outer worlds. Peirce does not directly address that issue here, but we have already seen (e.g., in selections 1 and 3) that at least *some* of the mathematician's structures are (or could be) found in Nature; otherwise the mathematician would be of no use to the scientist. Peirce need not hold that all of those structures could be found in the outer world, though epistemic availability through diagrams arguably implies some kind of physical realizability. In any case it is clear from the example of focal lengths in selection 10 that the scientific understanding of Nature involves, among other things, the recognition of mathematical structures there. And Peirce has strong systematic reasons for expecting the respective Thirdnesses of the inner and outer worlds to overlap. The mutual adaptation of Reason and Nature is a guiding conviction of his mature philosophy; without it abduction could not have the epistemic force that he ascribes to it.[1]

Peirce advances the general semiotic principle that every assertion involves all three basic kinds of signs,[2] and then proceeds to review a number of putative logical and mathematical counterexamples. The mathematical examples are bookended by two logical ones: definition, which does not obviously involve indices, and syllogism, which is alleged to have no need of icons because it makes no appeals to intuition. Peirce's response to the first example is that the purpose of definition is to clarify a partially indistinct concept; indeed, the subject of a definition is "not the concept in its pure ideal being but . . . the indistinct apprehension of it," and we can refer to this apprehension only by means of indices. His response to the second is that even in simple and familiar examples of syllogistic reasoning the conclusion does not just restate information already found in the premises; in deriving the conclusion we must "compound relations" and then infer the relation asserted in the conclusion from the compound one. This inference rests on observation of an icon which is a sign of the intermediate compound relation. Here Peirce relies heavily on another semiotic principle, which he does not state here, namely, that the relational elements of propositions are iconic (Peirce 1903h, E2.277).

The main mathematical examples come from arithmetic and algebra, where it is indices whose necessity seems most questionable. Geometrical examples come in, too, interwoven around the main thread of the argument. The labels on distinguished points in a geometrical diagram are a favorite Peircean illustration of indices, and icons are held to be vital to the communication of the intended interpretation of such geometrical terms as 'straight'. Indices figure in the algebraic examples in a number of ways. Propositions of applied mathematics require indices because they make reference to objects of experience. Pure mathematical assertions containing quantifiers have an indexical element as a matter of semiotic principle: Peirce's proto-game-theoretic semantics for the quantifiers interprets quantified sentences as an assertion of the utterer's ability, or a challenge to the interpreter, to select individuals of an appropriate sort.[3] Distinguished objects within a mathematical structure can only *be* distinguished, Peirce argues, by means of indices. Most interestingly, and most problematically, he maintains that indices are indispensable to such assertions because the "system of quantity" underlying them can only be indicated. This immediately raises the question, which Peirce neither poses nor answers here, of *how* we can indicate an abstract structure like that of the natural numbers. This is no less than the notoriously intractable problem of mathematical reference. The epistemology of structure set forth in selection 10, and explored in the final pages of the introduction, may point toward Peirce's solution.

Art. 2. The hypothesis of the mathematician is always the conception of a system of relations. In order that they may be reasoned about mathematically, these relations must be conceived as embodied in some kind of objects; but the character of the objects, apart from the relations, is utterly immaterial. They are always made as bare, skeleton-like, or diagrammatic as possible. With mathematicians not born blind, they are always visual objects of the simplest kind, such as dots, or lines, or letters, and the like. The mathematician often passes from one mode of embodiment to another. Such a change is no change in the hypothesis but only in the diagrammatic embodiment of the hypothesis. The hypothesis itself consists in the system of relations alone.

Art. 3. Every special problem has its special hypothesis, in the relations of which [there] are sundry exceptions or defects of uniformity which might be removed without altering the general character of those relations. The more general hypothesis which results from omitting those exceptions is common to all the problems of the same branch of mathematics. For example, if the problem relates to the intersections of two conics, the points of the plane which are in the conics are conceived as exceptional, that is as being in so far different from other points of the plane. This is the special hypothesis. The general hypothesis is that of points upon a plane, or other *locus in quo*.[4]

Art. 4. The diagrams in which the hypotheses are embodied are of two kinds. In the one kind the parts of the diagram are seen in the visual image to have the relations supposed. In the other kind of diagrams, the parts have shapes to which conventions or "rules" are attached, by means of which the supposed relations are attributed, or imputed, to the parts of the diagrams. Geometrical figures are diagrams of the inherential kind, while algebraical formulae are diagrams of the imputations kind.

Art. 5. Probable reasoning does not, properly speaking, apply to mathematical hypotheses; because probability is a ratio of frequency *in the general course of experience.* Now of the substance of an hypothesis, which has only an ideal being, there is no experience, and hence no course of experience. But probable reasoning may in certain senses be used in mathematics. For instance, when we add up a column twice and infer that there is no mistake because the results agree, this is genuine probable reasoning, referring to a course of experience. Namely, we infer that all sufficiently careful summations of the same column will give the same result as long as the same hypothesis, that is, the system of whole numbers, subsists. There is also a sort of modified probable reasoning when we find a certain property holds for many numbers and conjecture that it holds for all numbers. Here the course of the numbers is substituted for the course of experience, and there is something closely analogous to probability. But such reasoning is excessively weak unless it is supported by reasoning of a different kind. The reason is that there is no such thing as selecting numbers entirely at random, for there are infinitely more that are greater than any that can be selected than less than those numbers; so that we have only a right to infer that the relation is likely to hold if the numbers are not too great. When we infer the law of a series from a few terms, the reasoning is different. We know for certain that the series has some law and if we also can see that the law cannot be very complicated, a few terms must suffice to exhibit it. In case we have no assurance of the comparative simplicity of the law, the reasoning becomes excessively treacherous.[5]

The regular reasoning of mathematics is strictly deductive [or] necessary. It is analogous to syllogism. But the usual analysis of syllogism is very insufficient. It depends essentially upon the observation of diagrams. Thus, by the rule of transposition we infer from $x + y = z$ that $x = z - y$. But for this purpose, it is necessary to *observe* that $z - y$ is the effect of subtracting y from z. It clearly cannot be inferred by syllogism, as syllogism is usually understood. For the term $z - y$ occurs in the conclusion without being contained in either of the two premises, namely, first, the rule of transposition, and secondly, the equation $x + y = z$. We can only recognize that because $y - y = 0$, therefore $x + y - y = x + 0$ (even though the principle of association involved be admitted) except by observation; and we can make sure of it in no other way than that by which we make sure of the sum of a column of fifty figures.

Art. 6. For the purposes of the present memoir, the chief element of mathematics upon which attention has to be fixed is the hypotheses. The general hypothesis of a branch of mathematics supposes a system of individual objects related to one another in a simple and general way. The hypothesis of each special problem supposes other objects whose relations to one another are to be determined by assigning to each one of them one of the individual objects of the general hypothesis, so that the system of these latter objects serves as a middle term of comparison, for determining the relations of the objects of the special hypothesis to one another.

For example, the objects of the general hypothesis may be the points, or indivisible places, of space. The objects of the special hypothesis may be particles, or indivisible things in space. Then the relations of these particles to one another are determined by ascertaining the point of space occupied by each particle. Thus the places are middle terms of comparison for determining the relations of the things.

A multitude of individual objects in regular systematic relationship to one another, if it is employed as a middle term of comparison for determining the relations between other things, may be called a *scale of quantity*.

Thus, the system of real numbers is used as such a middle term of comparison; and so is the system of imaginary numbers. The last system has two dimensions; but that does not prevent mathematicians from calling it a system of *quantity*. Consequently, the system of points in tri-dimensional space ought equally to be called a scale of quantity. [. . .]

Art. 3. Although an exposition of the theory of deductive reasoning is beyond the province of this memoir, it will be necessary to take account of some features of it not noticed in the ordinary books, in order to explain why quantity plays so leading a part in mathematics.

Mathematical hypotheses are arbitrary creations of the mind. As such, their substance is not *experienced.* For an experience is the irresistible influence from without which an incident exerts upon the mind. An experience may be either cognitive or emotional; but by experience philosophers mean the aggregate of cognitive experiences. Life presents a course of experience; and that the like of which frequently happens in the general course of experience is said to be *probable*. In mathematics there is no variation of probability, except truth and falsehood; and no probable reasoning can properly be applied to the substance of a mathematical hypothesis or its consequences in themselves. We may, however, and can only assert any such consequence as probable; inasmuch as we may have committed an error in the deduction of it. For this deduction is an experiential phenomenon. Thus, we may cast up a long column of figures twice to assure ourselves no mistake has been committed.

In the operations of the mind three different kinds of elements are distinguishable by the independence of their intensities. The first kind embraces the feelings. By *feeling* is here meant that which is immediately present in consciousness. That is to say, it is wholly present in any one moment and endures without coming or going. Thus, the color of vermillion under a given degree of illumination is a feeling. It has its luminous intensity, its chromatic intensity, and its intensity of specific hue. Red is a more general feeling composed of vermillion and other reds taken together. By feeling is thus here meant the matter of consciousness.

The second element of consciousness is the consciousness of duality in the opposition, or over-against-ness, or reaction between subject and object, between *ego* and *non-ego*. The intensity of this element of consciousness is called the *vividness* of the idea. The reverse of vivid is called *dim*. There is a vividness due to the actuality or nearness of experience, which distinguishes perception from imagination, and a superposed vividness due to the mind's action in *attention;* and an idea which excites attention is said to be *interesting*. Thus, the memory of a great flash of lightening becomes, after years have elapsed, so dim that we roughly say it is not in consciousness at all, unless it be *recollected,* that is unless attention is focussed on it, when it becomes relatively vivid, although incomparably less so than on its first emergence. On the other hand, the tick of a watch, as one lies in the dark in a bed listening to it, is vivid in the highest degree, although as a feeling it is so faint that were it not for the aid of attention one could hardly say whether he had heard it or not. Other things being equal an intense feeling attracts more attention than a faint one, and is therefore more vivid. Vividness is suddenly altered in acts of sense and of will. When an outward object affects the senses, the vividness of a cluster of feelings is suddenly raised, practically from nothing to an extremely high grade. When the will acts upon an external object, that vividness of the idea which we call desire is suddenly lowered.

The third element of consciousness is the consciousness of the clustering together of ideas into sets. It shows itself intensively in the superior *suggestiveness* of certain ideas. Ideas are forever clustering more and more. The action is selective. An idea is more suggestive of one idea than of another.

But other things being equal, vivid ideas cluster and suggest others more energetically than dim ideas. A given cluster draws first those ideas for which it has the most energetic attraction. That is to say, all ideas move toward it, but some come into close union with it sooner than others. The total vividness is increased when the clustering tends toward a reaction between the inner and outer worlds. For this is no more than to say that that which concerns our desires is interesting. The total vividness is diminished when the clustering tends away from reaction between the inner and outer worlds. For the abstract is less vivid than the concrete. Hence, as a general rule, the stronger attractions acting first, any given combination of ideas grows to a maximum vividness and then subsides. The relative vividness of the constituents is equalized in the clustering. The way in which this acts is this. A certain number of ideas, A_1, A_2, A_3, etc. have had their separate vividnesses much reduced by clustering. They attract a dim idea, B, and the vividness of the whole cluster of As is much reduced, while that of B is much increased. Then, this cluster of the As with B attracts C_1, C_2, C_3, etc. and has its vividness much reduced. The consequence is that the As are no longer discernible to ordinary attention, and the result seems to be composed of B and the Cs, alone. The cluster of As has past its climax of vividness.

The clustering of ideas is either due to an outward occult power or to an inward one. That it is due to some occult power is plain from this that the ideas although they are in our own minds and thus normally subject to our will, cluster in spite of our will, and that in certain regular ways. This is a sound argument that some power not ourselves does that which ordinarily we ourselves do. But it is occult in this sense, that nothing more about it can be learned by mere observation of these phenomena. As to the distinction between the inner and outer worlds, it rests upon the phenomena of sense and will. There are some objects over which we find our control is all but absolute, while over all the rest it is all but nil. This distinction, though it is only one of degree, sunders the inner world from the outer by a great gulf. There are some ideas which become clustered together, because experience brings them together regularly; and consequently when one is called up, the rest of the set follow after. Subsequently, by the general rule of action explained above these latter remain vivid after the first has faded out. Hume calls this whole phenomenon *Association by Contiguity*.[6] But it is to be observed that the contiguity *consists* in ideas being brought together in experience, and is not the *cause* of it. That cause is that occult power acting like our wills, though with far greater might, which lies behind experience, and which the old philosophers called *Nature*. Every such action, when our attention is called to it, excites our wonder, and we ask the explanation of it. That is, we desire that the action of outward Nature should be rationalized, or reconciled to our own modes of action. Other ideas cluster together of themselves independently of experience. As in the other case, when one of the set is called up, the rest speedily follow, and some of these remain vivid after the first has faded out. This phenomenon is called by Hume *Association by Resemblance.* But as before, it is to be remarked that the resemblance *consists* in the ideas clustering together, (as scarlet and crimson, insist upon clustering together), and is not the *cause* of the clustering. That cause is an occult power which seems to lie behind the inward world just as Nature lies behind the outward world. It is often called *Reason.* Just as the positive sciences are founded upon the consistency of action of Nature, so mathematics is founded upon the consistency of action of reason. The action of Nature is a wonder to us; but that of Reason is not usually so. We are not surprised that Scarlet and Crimson should be alike. All that we demand of science is that it should show nature to be reasonable. Further than that we do not usually go. We seem to comprehend Reason. We flatter ourselves we grasp its very *noumenon.* But it is really as occult as Nature. It is only because its effects are for the most part familiar to us from infancy that they are not surprising. For when we come upon some property of numbers which is new to us, although it can spring from nothing but Reason, we are greatly surprised and begin to talk of the *Mystery* of Numbers, as of something which it is desirable to explore or which is incomprehensible. What we here demand is the mode of evolution of the action of Reason.

I have chosen a psychological form for the above statements; but their truth is not limited to mind. In the outer world there are also qualities, that is, Aristotle's forms, which are substantially of the nature of feelings, as these are defined above. The actualization of qualities consists in their action as forces; and forces are reactions between pairs, like the second element of consciousness. To the clustering of ideas corresponds the continual clustering of more and more causal factors, producing an evolution of more and more complex forms which appears in every kind of development. This is metaphysics: but it differs only from a scientific statement in resting not on special observations but on the ordinary observation of every man. However, I allow these few lines to it here only because it serves to give unity to the conception of deductive reasoning.

In the investigation of logic, the recognition of the three elements is the best light for our feet. Namely, it is necessary to recognize, first, unanalyzed qualities as in predicates of single subjects, secondly, dual relations, or predicates of pairs of subjects, and thirdly, mediation, as in plural relations, or predicates of sets of more than two subjects. The particular form of mediation, or clustering, which appears in deductive reasoning is the compounding of relations. Indeed, deductive reasoning may well be defined as the observation of the essential effects of compounding relations.

In order to comprehend the nature of mathematics, it is particularly useful to recognize the three elements as they appear in the three kinds of signs which have to be employed in logic. A sign, or representamen, involves a plural relation, for it may be defined as something in which an element of cognition is so embodied as to convey that cognition from the thought of the deliverer of the sign, in which that cognition was embodied, to the thought of the interpreter of the sign, in which that cognition is to be embodied. There are three ways in which such embodiment may be effected so as to serve the purpose. First, if the element of cognition to be conveyed is an unanalyzed abstract idea, or feeling, the only way is to present an icon, that is, a copy or exemplar of it. It is impossible to give a notion of redness except by exhibiting it. A proposition of geometry may be stated in general terms as in the usual enunciation of a theorem. But since definition must end in terms undefined, that statement cannot be understood until the student translates it into terms of a diagram which exhibits the relations meant. It is futile to attempt to convey an idea of a straight line, for instance, otherwise than by showing something so nearly like it as to suggest it. Second, if the element of cognition to be conveyed is an individual object of experience whose identity is determined by continuity of space and time, it only can be conveyed by first making it (if it be not already) an experience common to deliverer and interpreter, and second by forcibly directing the attention of the interpreter to it. A sign which fulfils this function, like the pointing of a finger, or an exclamation like "hi!" "see!", or a pronoun like "I," "you," "this," or a proper name, as yard, kilogram, or a word relative to the position of the interpreter, as "on high," "clockwise," "yesterday," I have called an index. Of this nature are the letters on a geometrical diagram (which are indispensible, substantially.) Third, if the element of cognition to be conveyed is predicative, that is, represents something to be true, which always consists in an identification or recognition of the object of a given index as an object of a given icon, a third kind of sign is manifestly required. For an icon only conveys a free dream without any forcefulness and an index only forces the attention without any general, rational, or qualitative element. But what has now to be expressed is a compulsion upon the mind which is conditional, a compulsion, not to think of the object of a given index, but *if* the object of that index be thought of, to think of it as an object of given icon. This sign, therefore, has to represent the clustering upon thought of experience. I call it a *symbol;* for it throws thoughts together.

It is evident that in every language whatsoever whether it be of the nature of speech, or writing, or what, if an assertion is made the signs of all those three kinds must be used. Thus, in order to express that fairies are pretty little creatures, I must have an index to show what experience I am speaking of, not here the things revealed in the great experience of the cosmos, but the things in a certain folklore which is not to be described simply,—because were it merely described the assertion would become an identical one, that a folklore that made fairies pretty *does* make them pretty, which is merely in the form of an assertion, but is empty of the matter of assertion,—but has to be *indicated,* as *that* folklore with which we have all been so familiar in childhood. The *icon,* which is the photograph of which the index is the legend, is here a complex one. It pictures an alternative, namely, on the one hand, that the object designated is not a fairy, on the other hand, that that object is a pretty creature. The symbol says that if the interpreter takes an object of that folklore, that alternative icon applies to it, in one of its two alternatives.

It would seem, at first blush, as if this analysis of assertion failed in the case of a definition, or logical analysis. For such an assertion does not relate to experience but merely to the essence of ideas. But were the definitum, or subject of a definition, to convey no other idea than exactly that conveyed by the definition, or analysis in the predicate, it would not be an assertion, except in form. In that case, no definition would be needed. But a conception may be familiar, and in that sense *clear,* without being *distinct,* that is, without its abstract constituents being recognized. Now that which is familiar *is* an object of experience, and in so far as a concept is recognized, not by logical apprehension of its elements, but by certain familiar marks, among the chief of which is its association with the word that signifies it, those marks are indices, not indeed of the concept in its pure ideal being, but of the indistinct apprehension of it. Now, the real subject of a definition is the indistinct apprehension, concerning which the proposition conveys positive information, namely, that the analysis of the definition applies where it applies.

But this explanation does not apply to the mathematical proposition, which is likewise an assertion about purely ideal objects. It cannot here be said that the subject is, in any measure, indistinct. Take the assertion, "Seven is a prime number." How can it be said that any index is required here?

This proposition, like any other mathematical proposition, may be understood in several ways. It may be understood as a proposition about collections, meaning that a collection of seven cannot be broken up into a plurality of equal pluralities. Or, it may be understood of the scale of number itself. Observe that we call it *the* scale of number, implying that it is a single collection. In this sense the assertion is that that individual of this collection the proper name of which is 7 cannot be produced by the multiplication of any two other individuals of the collection. In this sense, the assertion plainly relates to individuals, of one of which 7 is the proper name. Now a proper name is an index. Now, this sense is the pure mathematical sense. When we say a plural of seven units cannot be broken up into a plural of equal plurals, this is applied mathematics. Still, it may be said there is no index here. But the answer is that to conceive of seven units, it is necessary to distinguish them in thought otherwise than by their differences in quality. For if we speak of a red, an orange, a yellow, a green, a blue, a cyan, and a violet thing, meaning that there are seven in all, we must mean that there is just one of each color; and then those colors become proper names. Now proper names are indices. For mere specification of characters, however far continued, will only make smaller and smaller species, never necessarily an individual.

Nor is this a peculiarity of discrete number. In every branch of mathematics, every proposition whatever, true or false, if taken in the sense of pure mathematics relates to a system of quantity, under which head I include a space, or in Cayley's phrase the *locus in quo,* which is an individual thing, and so can only be designated by an index, and furthermore there is taken, or directed to be taken, a special determination of it, which owing to the indistinctness of its specification, can likewise only be designated by an index. For example, let the proposition be, "Every algebraic equation has a real or imaginary root." This refers to *the* field of imaginary quantity, which, as the definite article shows, is taken as an individual system. This has to be designated by an index, although certain propositions are true of it; so that certain icons are applicable to it, too. Besides this index another is required. For the assertion is that, taking any function of one variable in this system, it will be found either to be non-algebraic or to have a zero. This function is *any* one. It is not described. It must be in some way identified, whatever one is selected. It is to be selected at the will of the interpreter. This act of will has to be represented; and an act of will can only be represented by an index, because of its element of arbitrary force. So with the proposition, "Some equation has no real root." Here the function is to be selected at the will of the deliverer; and for the same reason as before an index is required. If the proposition is taken in the sense of applied mathematics, still more, if possible are indices required, because it now refers to some familiar experience. Thus, if we say in this sense that any two straight lines that intersect lie in a common plane, what we mean by straight lines no formal definition can possibly say. They are two objects selected at the will of the interpreter out of the entire multitude of objects familiar to experience under the name of straight lines, passing through one of the indescribable objects familiar to experience as points, the particular point to be selected at the will of the interpreter.

Thus every assertion requires indices whether it refers to real external experience or only to an inward experience of objects of the mind's creation.

It might also be objected that the enunciation of a geometrical theorem involves no *icon,* because there is nothing in the sounds (if it is spoken) or in the shapes of the characters (if it is written) which is like the geometrical object. But the answer to that is, that though no icons are used in the outward expression, the meanings of the words cannot be understood without *icons* in the minds of both deliverer and interpreter.

It is hardly supposable that anybody should object that no symbol should be required, although in many languages the "substantive" verb, as it is erroneously called, is frequently omitted, as in the dialect of English employed for telegrams. The reply obviously would be that the symbol is mentally supplied both by deliverer and interpreter.

The traditional conception of syllogism, which is accepted by all who are somewhat imbued with the traditional logic, without having studied it very critically, which class of minds embraces the greater part of educated men, as well as the larger proportion of philosophers, undoubtedly is that deductive reasoning is performed "symbolically," as Leibniz says, without any aid from "intuition," that is, from icons.[7] But the truth of the matter is that in such reasoning the icon is the most vital element. For such reasoning always consists in stating a complex relation and then observing that that relation involves another relation, which is said to be inferred, concluded, or deduced. In all but the very simplest cases, deductive reasoning consists in compounding relations; and this compound relation has to be expressed in a term which was not contained in either premise. Hence, syllogism, as ordinarily understood, which has no terms in its conclusion not found in its premises, is inadequate to the representation of such reasoning. Even the simplest cases are substantially of this kind. That is to say, the conclusion states a relation not stated in either premise. Syllogism is not so simple an act as is supposed. Given the premises

> All men are mortal,
>
> Enoch is a man,

the first step is to colligate the premises, which requires wit, and to infer the single proposition

> Enoch is a man and all men are mortal.

Another step (which we may here regard as simple) infers from this,

> Enoch is a man that is mortal;

whence,

> Enoch is something that is mortal.

Finally, we conclude

> Enoch is mortal.

8

Sketch of Dichotomic Mathematics

[Peirce 1903(?)] Like selection 4, this text is a philosophical preliminary to a treatment of "the simplest possible mathematics." It is Peirce's explanation to the reader of the terminology he will use to organize his presentation. That terminology is taken largely from Euclid's *Elements,* and Peirce's explanations are partly a commentary on the traditional vocabulary, partly an exposition of his own theory of mathematical knowledge and reasoning.

In accordance with his views on the ethics of scientific terminology, Peirce promises to defer to tradition in his usage of 'definition', 'postulate', and 'axiom', while reserving the right to resolve the vagueness of the remaining terms as best suits his purposes. He does a creditable job of keeping his promise with the first two terms, though the reader may find it useful to compare what he says with Heath's account of Euclidean first principles in the introduction to his translation of the *Elements* (Heath 1926, 117–124). What Peirce says about definitions here is also worth comparing with what he says about them in selection 7. When he gets to axioms, however, he cannot resist the temptation to improve upon tradition, his promise notwithstanding. For Aristotle an axiom is an indemonstrable starting point, a "common notion," which can be invoked in any scientific argument: "a proposition," as Peirce puts it, "which any learner would recognize as true." Peirce immediately complicates this straightforward statement in a revealing subordinate clause: "and which, since no doubt was entertained about it, could not but be assumed to be true." He is distancing himself here from Aristotle's assumption that axioms are in fact true; a few lines later he will give an example of a Euclidean common notion—"that the whole is greater than its part"—which turns out to be false. A further complication stems from the mathematician's exclusive concern with hypothetical states of things; how can we have any indemonstrable knowledge of such a state, which is not already contained in the postulates that define it? Axioms thus seem to be otiose in mathematics.

Peirce's own account of axioms attempts to address both of these complications: an axiom is a proposition which is "immediately evident" *given* the postulates; we come to accept it because "the enunciation of the definitions and postulates puts us into position to observe other facts than those which they assert." Peirce argues that such observations play an essential role in the comprehension of any demonstrative argument: this is his explanation of Aristotle's remark that there is no need to refer to axioms in demonstrations. We will see that for Peirce axioms play an essential role in theorematic reasoning. So they are far from otiose for him, but at the same time they introduce a very real risk of error. Like many more recent writers on the foundations of mathematics, Peirce sees axioms as playing a largely organizational role in mathematical knowledge; they are of "no other use than that of showing what the foundations of theorems are: they render them neither more certain nor more clear."

He sounds another forward-looking note when he supplements the traditional vocabulary with 'convention'. He notes that it is not always "easy to distinguish between a fundamental assumption concerning the subject of a deductive reasoning, and a fundamental convention respecting the general signs employed in that reasoning." Peirce believed that the results of Cayley and Klein outlined in the headnote to selection 16 (119–120) implied that those Euclidean postulates whose effect is to lay down a system of measurement are in fact conventions; in keeping with his claim, in that selection, that topology is "the only mathematics of pure Space that is possible" (122), he states here that "the only true postulates of geometry are the topical postulates."

The distinction between conventions and postulates, then, can be difficult to draw in practice. Peirce does not go so far as to suggest that there might be cases in which it cannot be drawn at all in a non-arbitrary way. But his semiotic explanation of the difficulty makes it look like a fairly deep-seated fact about mathematical reasoning: the distinction is so delicate because "the primary subject of a course of deduction is . . . of the nature of a general sign," whose operations are largely a matter of convention. In selection 10 the scope of convention will be enlarged, and the delicacy of this distinction thereby intensified.

As one would expect, Peirce's treatment of the terms 'corollary' and 'theorem' turns mainly on his distinction between corollarial and theorematic reasoning, which we have already encountered in the roughly contemporaneous selection 4. The two explanations of the distinction are well worth comparing. The main difference between them, as far as corollaries go, is that here Peirce insists that the proof of a corollary should rest only on postulates and definitions, and not refer to other corollaries.

In this definition of 'theorem' Peirce stresses that a theorem asserts the *impossibility* of a certain result given the theorem hypothesis. Why impossibility and not just plain necessity? Evidently because he wishes to highlight the modal difference between the end product and the process of theorematic reasoning, and also the difference between theorems and problems (see below). The auxiliary constructions that make the process theorematic must be *possible;* axioms come into play here, as does the logic of "substantive possibility" (see selection 12). Peirce suggests that the distinction between theorems and corollaries, like that between postulates and conventions, is a delicate one: he gives an explicit definition of equality that goes far towards turning the *Pons Asinorum* into a corollary. Further light is cast on the distinction by a canonical form for proofs of theorems. In laying out the canon Peirce touches also on the semiotics of proof, especially on the respective roles of icons and indices.

Peirce's definition of 'problem' marks his apparently most substantial departure from the tradition, which had connected theorems with knowledge and problems with action. The founder of pragmatism is of course disinclined to sever knowledge from practice too neatly; in any case the traditional account masks the really important, modal difference between theorems and problems, namely that theorems deny, and problems affirm, existence or possibility.

Scholium. The purposes of this sketch are

1st, to put the reader into a position to understand what dichotomic mathematics is, why it is so called, and that it is the simplest possible mathematics and the foundation of all other mathematics;

2nd, to develope its main propositions and methods;

3rd, to exhibit, in mathematical style the analysis of its foundations into their simplest formal elements, making clear the different categories of conceptions involved, and indicating the logical doctrine with which the subject is connected as well as can be done while strictly avoiding all direct logical discussion.

I shall prefix to each article a descriptive heading, inventing terms for the purpose when necessary, but as far as possible availing myself of those found in editions of Euclid's Elements; viz., Definition, Postulate, Axiom, Theorem and Demonstration, Problem and Solution, Corollary, and Scholium. The first three of these are technical terms whose signification could not be changed without a violation of the ethics of terminology.[1] To the others, I consider myself at liberty to attach definite meanings not in violation of their present vague meanings.

A *Definition* is either *Nominal* or *Real.* A nominal definition merely explains the meaning of a term which is adopted for convenience. I shall not make separate articles for such definitions nor state them formally. For they do not affect the course of development of the thought. A *Real Definition* analyzes a conception.[2] As Aristotle well says (and his authority is well-nigh absolute upon a question of logical terminology,) a definition asserts the existence of nothing.[3] A definition would consist of two members, of which the first should declare that any object to which the *definitum*, or defined term, should be applicable would possess the characters involved in the definition; while the second should declare that to any object which should possess those characters the definitum would be applicable. And any proposition consisting of two members of this description and really contributing to the development of the thought would be a Real Definition.

The term *Postulate* is carefully defined by Aristotle, whose acquaintance with the language of mathematics is guaranteed by the fact that he was long in Plato's school at the Academy, taken in connection with his known intellectual character; and Heiberg's recension of the Elements renders it manifest that Euclid used the term in the same sense.[4] This meaning has become established in the English language, which has been peculiarly fortunate in inheriting much of the medieval scholastic precision, and is by far the best of the modern languages for the purposes of logic. To this meaning, therefore, the ethics of terminology peremptorily commands us to adhere. Long after Euclid a quite different sense came to be attached by some writers to the word. Namely, they took it to mean an indemonstrable practical proposition. The influence of Christian Wolff has caused the word *Postulat,* in German, to be generally used to signify the mental act of adhesion to an indemonstrable practical proposition.[5] This usage is very abusive. Generally, the German treatment of logical terms is bad. Little of medieval precision of logic is traditional in Germany. I have read a great many German nineteenth-century treatises on logic;—probably fifty. But I have never yet found a single one, not even Schroeder's, which did not contain some unquestionable logical fallacy.[6] It would be difficult to find a single instance of such a phenomenon in an English book by a sane author. German logical terminology was early vitiated by the national tendency towards subjectivism. It was further corrupted by the nominalism of Leibniz. With Hegel came chaos, and all the restraints of terminological ethics seemed to give way.

A *Postulate* would be a proposition necessary as a premiss for a course of deductive reasoning and predicating a contingent character of the hypothetical subject of that course of reasoning; and any proposition of that description would be a Postulate.

It seems, at first sight, as if it must be easy to distinguish between a fundamental assumption concerning the subject of a deductive reasoning, and a fundamental convention respecting the general signs employed in that reasoning. But the primary subject of a course of deduction is itself of the nature of a general sign, since there can be no necessary reasoning about real things except so far as it is assumed that certain general signs are applicable to them. It therefore becomes a delicate matter to draw a just line between the subject of the reasoning and the machinery of the reasoning. The subject of geometry will afford an illustration. Be it observed, in the first place, that it does not fall within the province of the mathematician to determine whether it is a fact that a plane has a line at infinity, as projective geometry assumes, or a point at infinity, as the theory of functions assumes, or whether or not any postulate of geometry is precisely true in fact.[7] Consequently, his whole course of reasoning relates to an idea, or representation, or symbol. But everybody would agree that the description of a system of coördinates was a convention and not a postulate concerning pure space. A system of measurement, however, is of the same nature as a system of coördinates. For to the same space may be applied at will elliptical, hyperbolical, or parabolic measurement.[8] If, therefore, the description of a system of coordinates is a convention and not a postulate, which is unquestionable, the same must be said of every proposition determining the system of measurement employed. Yet this is the nature of Euclid's fourth and fifth postulates. The truth is that the only true postulates of geometry are the topical postulates.[9] I shall head every article describing the machinery to be employed in the discussion as a *Convention.*

A *Convention* would be a proposition concerning a subject which we imagine to exist as an aid in drawing conclusions concerning our main subject; and any such proposition would be a Convention.

The *Axioms* are called by Euclid common notions, κοιναὶ ἔννοιαι; and Aristotle occasionally applies the same designation to them. A little later than Euclid, the term common notion, κοινὴ ἔννοια, became a technical term with the stoics; and they undoubtedly meant by the appellation, κοινή, that they were common to all men, a part of the "common sense" from which the human mind cannot escape. It was the same idea, though less distinctly apprehended, perhaps, which Aristotle had in view. An axiom was a proposition which any learner would recognize as true, and which, since no doubt was entertained about it, could not but be assumed to be really true.[10] But considering that pure mathematics or any pure deduction can deal only with a purely hypothetical state of things which is fully described, so far as it is supposed at all, in the postulates, how can we know anything about it, or what would it mean to say that anything was true of it, beyond what can be deduced from the postulates? What sort of a proposition could an axiom be? My reply to this question may be embodied in a definition, as follows:

Any *Axiom* would be a proposition not deducible from anything asserted in the definitions and postulates, but immediately evident in view of the facts that these have been laid down and are true.

In order to make the meaning of this clear, suppose that I make a judgment, that is, that I say to myself 'S is P.' Some German logicians tell us that this is logically identical with saying to myself, 'I know that S is P.'[11] But this is not so. In the former proposition nothing whatever is said about me or my knowledge. And if the two propositions were identical, their denials would be identical; but the denial of one is 'S is not P,' while that of the other is the very different proposition 'I do not know that S is P.' But the truth is that, granting that S is P, then by the observed fact that I have said to myself that S is P, I perceive that it is true that I know that S is P. Thus, the enunciation of the definitions and postulates puts us into position to observe other facts than those which they assert; and I suppose that an axiom is of that nature, because there is no other nature that it can have. Besides, this is confirmed by a rather striking remark of Aristotle to the effect that it is not necessary in a demonstration to refer to an axiom.[12] Why should it not be necessary to refer to a premiss? My definition affords the explanation. It is because at every step of a demonstration it is necessary to make similar observations in order to apprehend the force of the reasoning. A stupid person may admit almost in one breath the two premisses of a syllogism, and think them, too, and yet not see the truth of the conclusion until it is pointed out. To do that it is necessary to observe that that very character, M, which belongs to S, is the same character that carries with it P. The introduction of an axiom, therefore, spares the reader no difficulty, since to see the application of it is quite as difficult as to see the consequence for which the axiom was supposed to afford a stepping-stone. Axioms are, in truth, of no other use than that of showing what the foundations of theorems are: they render them neither more certain nor more clear. In a backhanded way, indeed, a good logician will make use of them; for where an axiom is cited there is a likely place to find a fallacy. Euclid in his first book uses his axiom that the whole is greater than its part seven times. In the 6th and 26th propositions it was not needed; and these propositions are strictly true. The 7th, 16th, 18th, 20th and 24th propositions, where it is used, are, in consequence of this, untrue for triangles whose sides pass through infinity, or are untrue for elliptic space, or both.[13] About as true as the axiom about the whole and part are those about equals subtracted from equals giving equals, etc., as if infinite lines cut from infinite lines would necessarily give equal finite remainders. Yet I have known a professed modern mathematician to cite this axiom to prove that infinitesimals are strictly zero. Since he wrote this to me, a logician, he seems not to have dreamed that the argument was open to any objection. He voluntarily gave me permission to make any use of his letter; but I have no taste for seeing respectable people held up to ridicule.

Any *Corollary,* (as I shall use the term,) would be a proposition deduced directly from propositions already established without the use of any other construction than one necessarily suggested in apprehending the enunciation of the proposition; and any such proposition would be a Corollary.

The proof of a corollary should not only make it evident, but should show clearly upon what it depends. The proof should, therefore, never cite another corollary as premiss but should be drawn from postulates and definitions, as far as this can be done without a special construction.

Any *Theorem* (as I shall use this term) would be a proposition pronouncing, in effect, that were a general condition which it describes fulfilled, a certain result which it describes in a general way, except so far as it may refer to some object or set of objects supposed in the condition, will be impossible, this proposition being capable of demonstration from propositions previously established, but not without imagining something more than what the condition supposes to exist; and any such proposition would be a Theorem.

For example, the *Pons Asinorum*[14] may be proved by first proving that a rigid triangle may be exactly superposed on the isosceles triangle, and that it may be turned over and reapplied to the same triangle. But since the enunciation of the *Pons* says nothing about such a thing; and since the *Pons* cannot be demonstrated without some such hypothesis, and since, moreover, the *Pons* does not pronounce anything to exist or to be possible, but only pronounces the inequality of the basal angles of an isosceles triangle to be *impossible,* it is a theorem. Perhaps, however, spatial *equality* could not be better defined than by saying that were space to be filled with a body called "rigid," which should be capable of continuous displacement, freely interpenetrating all other bodies, its rigidity consisting of the facts, first, that every film (or bounding *part* between *portions*) of it which at any instant (or absolutely determinate state of its displacement) should fully occupy any one of the surfaces of a certain continuous, unlimited family of fixed and topically non-singular surfaces of which any three should have one, and only one, point in common, would at *any* instant fully occupy one of those surfaces, and second, that there should be one of those surfaces that should at all instants be fully occupied by the same film, then any two places, or parts of space, which could be fully occupied at different instants by one and same part of that rigid body would be *equal;* and any two parts of space that would be equal would be capable of being so occupied, provided the rigid body had all the freedom of continuous displacement that was consistent with its rigidity.[15] Were such a definition admitted, we may admit that the idea of a rigid triangle being turned over and reapplied to the fixed isosceles triangle would be so nearly suggested in the enunciation that the proposition might well be called a corollary. Perhaps when any branch of mathematics is worked up into its most perfect form all its theorems will be converted into corollaries.[16] But it seems to be the business of mathematicians to discover new theorems, leaving the grinding of them down into corollaries to the logician.

There are propositions whose proofs are accomplished by means of constructions so well understood to be called for in proving propositions of the classes to which those propositions belong, that it is a delicate matter to determine whether they are best called corollaries or theorems. But though this were inherently impossible in some cases, my distinction between a corollary and a theorem would not thereby be proved ill-founded.

Sundry advantages not to be despised recommend the adoption of a canonical form of statement for all demonstrations of theorems, the conformity to the canon being subject in each case to such limitation as good sense may impose. The following is the canon which I shall, in that sense, adopt:

1. The theorem having been enunciated in that form which is most convenient for conveying it in mind may, in the first place, be put by logical transformation, into such form as is most convenient for the purpose of demonstration.

2. Supposing the enunciation to be still in general terms, an icon, or diagram must next be created, representing the condition of the theorem, in the statement to which it has been brought in 1. At the same time, indices (usually, letters) must be attached to those parts of the icon which are to be made objects of attention in the demonstration, for the purpose of identifying them.

3. Next we must state the *ecthesis,*[17] which is that proposition, which, in order to prove the theorem, it will manifestly be necessary and sufficient to show will be true of the icon created and of every equivalent icon.

4. Such additions to the icon as may be needful must now be created and supplied with indices.

5. It must be proved that these additions are possible. If the postulates (with the aid of previously solved problems) expressly render them so, this part of the demonstration will involve no peculiar difficulty. Even when this is not the case the possibility may be axiomatically evident. Otherwise it can only be proved by means of analogy with some experience; and to render the sufficiency of this analogy evident will be difficult. The logic of substantive possibility will be of assistance here.

6. It will be convenient at this point to make all the applications of previous propositions that are to be made to parts of the augmented icon. The several immediate consequences being set down and numbered, or otherwise indicated. Of course, the interpreter of the demonstration will have to observe for himself that the previous propositions do apply; but unobvious cases should be elucidated by the demonstrator.

7. The demonstrator will now call attention to logical relations between certain of the numbered propositions which lead to new propositions which will be numbered and added to the list, until the ecthesis is reproduced.

8. When this proposition is reached, attention will be drawn to its identity with the ecthesis by means of the letters Q.E.D. or something else equivalent to Euclid's phrase, ὅπερ ἔδει δεῖξαι. Euclid frequently employs the same phrase when he reaches the general enunciation itself without any ecthesis. But it would be more correct in such a case to vary the phrase, since the logical situation is not the same. I shall say "which is our theorem," or, since a repetition of the enunciation is needless and fatiguing, "whence, or from such and such propositions, our theorem directly follows."

If I wish to employ the *reductio ad absurdum,* the first article of the demonstration will consist in throwing the theorem into the form, "To suppose so and so leads to a contradiction," and my ecthesis will state definitely what contradiction it is that I shall find it convenient to bring out. Since a hypothesis which involves contradiction may be shown to be contradictory in any feature that may be chosen, the ecthesis in this case is determined by convenience.

Any *Problem* (as I shall use the term) would be a proposition pronouncing, in effect, that under circumstances known to exist or to be possible, a certain sort of result, described partly, at least, in a general way, exists or is possible, this vague proposition being usually, though not essentially capable of definition, showing under what general conditions it always will be realized and at any rate, being capable of demonstration from propositions previously established, but only by means of a hypothesis not immediately suggested in the vague enunciation; and any proposition of that description would be a Problem.

This definition certainly ventures to break somewhat with ordinary usage (for I do not think the word 'problem', in its mathematico-logical sense, can be called a term of art,) and appears to do so more than it does. It is true that as commonly used, it is supposed to describe a thing to be *done.* Every proposition has its practical aspect. If it means anything it will, on some possible occasion, determine the conduct of the person who accepts it. Without speaking of its acceptance, every proposition whatsoever, although it has no real existence but only a *being represented,* causes practical, even physical, facts. All that is made evident by the study which I call *speculative rhetoric.*[18] But I do not think that the practical aspect of propositions is pertinent to the designation of classes of mathematical propositions. For example, Euclid's first proposition ought to read that upon any terminated right line as base some equilateral triangle is possible. He first makes this definite by describing precisely how the triangle is defined. There can be a circle of any centre and radius. Then there can be two circles with centres at the two ends of the line and with the length of the line as radius. Had Euclid looked at his proposition in this light, he could not well have failed to perceive the necessity for proving that the two circumferences (which, as usual, he calls circles) will intersect. He would have proved this by the definitions of *figure* and of *circle,* and by his third postulate.[19]

So far, the difference between the ordinary acception of the term *problem* and that which I propose is trifling. But it frequently happens that a problem, or the vague statement that an object of a certain description exists, such as, a real root of the equation of an algebraic polynomial of odd degree involving but one unknown, while it still remains impossible to give an exact general description applicable to every such thing and nothing else; which would constitute, as I should say, a *solution* of the problem. The distinction between such a proposition affirming existence or possibility, and a theorem, or proposition denying existence or possibility, is the most important of all logical distinctions between propositions; while the distinction between *knowing* and *ability to do* is a distinction quite irrelevant to logic. Considering further that the limitation of the word *problem* to practical matters is confined to geometry, while in every sense of the word the idea of vagueness attaches to it, encourages me to believe that I may venture, without blame, to use the word in the sense defined.

9

[Pragmatism and Mathematics]

[Peirce 1903c] Peirce wrote this selection as one of his 1903 Harvard lectures on pragmatism, but did not deliver it. Given his aims and audience, this was probably the right choice; but this is nonetheless one of his richest and most comprehensive treatments of the philosophy of mathematics.

The definition of mathematics with which Peirce begins is more or less the one developed in the foregoing selections: mathematics is a purely hypothetical science, not restricted to but heavily reliant upon quantity. Peirce's statement of the definition is followed by two assertions that set the agenda for the lecture as a whole: that mathematics is a science, and that pragmatism cannot pass philosophical muster if it is found to "conflict with this great fact." The historical remarks that follow are probably designed to lower the audience's resistance to the definition. Peirce tells them that they stand at a turning point in the history of the science, in the middle of a renewal of the Greek ideals of rigorous reasoning, "a vast reform . . . not completed . . . [which] has quite revolutionized our conception of mathematics." As always when Peirce expatiates on the nature of mathematics, diagrams are seen to play an essential role, though he toys here with the possibility that their indispensability may be only psychological. He rejects the suggestion that the observational character of diagrammatic reasoning blurs the boundary between deduction and induction, a stance he will reconsider in selection 10, written three years later.

The heart of the lecture is a discussion of mathematical ontology. Characteristically, Peirce introduces the question by way of mathematical reasoning, which in his view has never been adequately analyzed, not even in terms of his own "algebra of logic." His attempted analysis did at least, he claims, tackle higher forms of mathematics (which remain unspecified) than Dedekind, Mill, and Schröder had done. As a result it is only Peirce's analysis that brings out the importance of abstractions. The ensuing account of what abstractions are, and the arguments for their reality, have already received a good deal of attention in the introduction (xxxviii–xxxix). One argument not mentioned there, which is the most important of all within the context of this lecture, rests on "pragmatistic principles"; though he does not announce it as such, this looks like Peirce's reconciliation of pragmatism with the "great fact" that mathematics is a science.

The concluding paragraphs of the lecture are an introduction to "dichotomic mathematics." Peirce's approach here is very general: the algebra of logic is mentioned as an important variety of dichotomic mathematics, but the focus is not on logic but rather on the surprising wealth of mathematical structures that are available in the general dichotomic setting. No doubt this was intended as a bridge to the third lecture, on Peirce's triadic system of categories. But its chief interest, from the standpoint of the present lecture, is in the use of relations to generate abstractions of higher and higher orders. Though Peirce does not treat his relations extensionally, as we might do, as collections (e.g., of ordered pairs), we will see in subsequent selections (e.g., selection 21) that some years before this he had already developed a theory of collections importantly similar to our own iterative conception of set. Peirce was well aware, when he wrote this lecture, that abstraction packs a powerful ontological punch.

I feel that I must not waste time; and yet an investigation, in order to be really solid, must not confine itself too closely to a question set beforehand. Let us not set our thoughts on pragmatism but survey the whole ground and let the evidences for or against pragmatism or in favor of a modification of it come when they will without being teased.

Pure mathematics is the study of pure hypotheses regardless of any analogies that they may present to the state of our own universe. It would be wild to deny that there is such a science, as actively flourishing and progressive a science as any in the whole circle, if sciences can properly be said to form a circle. It certainly never would do to embrace pragmatism in any sense in which it should conflict with this great fact.

Mathematics, as everybody knows, is the most ancient of the sciences; that is, it was the first to attain a scientific condition. It shows today many traces of its ancient lineage, some of which are excellences while others are unfortunate inheritances. The Greeks were very fine reasoners. Throughout the XVIIIth century, the opinion prevailed among mathematicians that the strictness of Greek reasoning was unnecessary and stood in the way of advances in mathematics. But about the middle of the XIXth century it was found that in important respects the Greek understanding of geometry had been truer than that of the moderns. Gradually, beginning we may say with Cauchy early in the XIXth century a vast reform has been effected in the logic of mathematics which even yet is not completed. This has quite revolutionized our conception of what mathematics is, and of many of the objects with which it deals, as well as of the logical relations between the different branches and the logical procedure. We are now far above the Greeks; but pure mathematics as it exists today is a decidedly youthful science in such an immature state that any student of logical power may very likely be in possession of important *aperçus* that have not yet become common property.

That mathematical reasoning is by no means confined to quantity is now generally perceived, so that it now becomes an extremely interesting logical question why quantity should play so great a part in it. It is recognized that the main business, if not the only business, of mathematics is the study of pure hypotheses and their consequences, or, as some say, the study of the consequences of pure hypotheses only. The Greeks approached this conception without attaining to it.

In the procedure of *all* mathematics whatsoever, the observation of diagrams plays a great part. That this is true in geometry was shown, though rather vaguely, by Stuart Mill in his logic.[1] It is not so obvious that algebra makes use of the observation of visual images; but I do not think there ought to be any doubt of it. Arrays of letters are observed, although these are mixed with conventional signs with which we associate certain so called "rules," which are really permissions to make certain transformations. There is still some question how far the observation of imaginary, or artificial constructions, with experimentation upon them is logically essential to the procedure of mathematics, as to some extent it certainly is, even in the strictest Weierstrassian method,[2] and how far it is merely a psychological convenience. I have sometimes been tempted to think that mathematics differed from an ordinary inductive science hardly at all except for the circumstance that experimentation which in the positive sciences is so costly in money, time, and energy, is in mathematics performed with such facility that the highest inductive certainty is attained almost in the twinkling of an eye. But it is rash to go so far as this. The mathematician, unless he greatly deludes himself, the possibility of which must be considered, reaches conclusions which are at once enormously and very definitely general and yet, but for the possibility of mere blunders, are absolutely infallible. Anybody who fancies that inductive reasoning can achieve anything like this has not made a sufficient study of inductive reasoning.

Induction is no doubt generalization and mathematicians,—especially mathematicians of power,—are so vastly superior to all other men in their power of generalization, that this may be taken as their distinctive characteristic. When we are dealing with the real world cold water gets dashed upon any generalizing passion that is not well held in check. There are very few rules in natural science, if there are any at all, that will bear being extended to the most *extreme cases.* Even that invaluable rule that the sum of the angles of a plane triangle is equal to two right angles shows signs of breaking down when by the aid of photometric considerations and that of the numbers of stars of different brightness, we compare statistically that component of stellar proper motions that is due wholly to the real motions of the stars with that component that is partly due to the motion of the solar system.[3] But when we come to pure mathematics we not only do not avoid the extension of principles to extreme cases, but on the contrary that is one of the most valuable of mathematical methods. In regard to any ordinary function for example, if we only know for what values of the variable it becomes zero and for what values it becomes infinite, we only need to know a single finite value to know all there is to be known about it.[4]

No minute analysis of any piece of characteristically mathematical reasoning has ever appeared in print. There are numbers of attempts which profess to be successful. There have even been professed representations of the reasoning of whole books of Euclid in the forms of traditional syllogistic. But when you come to examine them, you find that the whole gist of the reasoning, every step in the progress of the thought which amounts to anything, instead of being analyzed logically is simply stated in the form of a premiss. The only attempts that are in any important degree exceptions to this are Mill's analysis of the *pons asinorum* which has its value,[5] but which relates to too slight a bit of reasoning to teach much and Dedekind's little book on the foundation of arithmetic with Schröder's restatement of it.[6] This is certainly very instructive work. Yet it is open to both the same criticisms. In the first place, the mathematics illustrated is, most of it, of too low an order to bring out in strong colors the real peculiarities of mathematical thought; and in addition to that, the real mathematical thinking is, after all, only stated in pretty much the old fashion of all mathematical writers, that of abridged hints. It is not really analyzed into its logical steps. Every now and then the intelligent reader will say, "I wonder how he got the idea of proceeding so and so." But it is just at these points that the fine mathematical thinking comes in. It is left undissected.

When I first got the general algebra of logic into smooth running order, by a method that has lain nearly twenty years in manuscript and which I have lately concluded that it is so impossible to get it printed that it had better be burned,—when I first found myself in possession of this machinery I promised myself that I should see the whole working of the mathematical reason unveiled directly. But when I came to try it, I found it was the same old story, except that more steps were now analyzed. About the same amount as in Dedekind's and Schröder's subsequent attempts. Between these steps there were unanalyzed parts which appeared more clearly in my representation than in theirs for the reason that I attempted to analyze a higher kind of mathematics. I was thus forced to the recognition in mathematics of a frequent recurrence of a peculiar kind of logical step, which when it is once explained is so very obvious that it seems wonderful that it should have escaped recognition so long. I could not make any exact statement of it without being led into technical developments that I desire to avoid and which are besides precluded by my being obliged to compress what I have to say into six lectures. But I can give you a general idea of what the step is so that you will be able by subsequently studying over any piece of mathematics to gain a tolerable notion of how it emerges in mathematics.

Let me at this point recall to your minds that the correlatives *abstract* and *concrete* are used in two very little connected senses in philosophy. In one sense, they differ little from *general* and *special* or *particular,* and are for that reason hardly indispensible terms; though that is the usual meaning in German, which is so to say pushed to an extreme in Hegel's use of the words. The other sense in which for example *hard* is concrete and *hardness* is abstract, is more usual in English than in other languages for the reason that English is more influenced by medieval terminology than other languages. This use of the words is fully as well authorized as the other if not more so; and this is the sense in which I shall exclusively employ the words. Hard is concrete[,] hardness abstract.

You remember the old satire which represents one of the old school of medical men,—one of that breed to whom medicine and logic seemed to be closely allied sciences,—who asked why opium puts people to sleep answers very sapiently 'because it has a dormitive virtue.' Instead of an explanation he simply transforms the premiss by the introduction of an *abstraction,* an abstract noun in place of a concrete predicate. It is a poignant satire, because everybody is supposed to know well enough that this transformation from a *concrete predicate* to an abstract noun in an oblique case, is a mere transformation of language that leaves the thought absolutely untouched. I knew this as well as everybody else until I had arrived at that point in my analysis of the reasoning of mathematics where I found that this despised juggle of abstraction is an essential part of almost every really helpful step in mathematics; and since then what I used to know so very clearly does not appear to be at all so. There are useful abstractions and there are comparatively idle ones; and that one about *dormitive virtue,* which was invented with a view to being as silly as one could, of course does not rank high among abstractions. Nevertheless, when one closely scrutinizes it or puts it under a magnifying glass,—one can detect something in it that is not pure nonsense. The statement that opium puts people to sleep may, I think, be understood as an induction from many cases in which we have tried the experiment of exhibiting this drug, and have found, that if the patient is not subjected to any cerebral excitement, a moderate dose is generally followed by drowsiness, and a heavy dose by a dangerous stupor. That is simply a generalization of experience and nothing more. But surely there must be some explanation of this fact. There must be something, say to fix our ideas, perhaps some relation between a part of the molecule of morphine or other constituent of opium which is so related to some part of the molecule of nerve-protoplasm as to make a compound not so subject to metabole as natural protoplasm. But then perhaps the explanation is something different from this. Something or other, however, there must be in opium[,] some peculiarity of it which if it were understood would explain our invariably observing that the exhibition of this drug is followed by sleep. That much we may assert with confidence; and it seems to me to be precisely this which is asserted in saying that opium has a dormitive virtue which explains its putting people to sleep. It is not an explanation; but it is good sound doctrine, namely that *something* in opium must explain the facts observed.

Thus you see that even in this example which was invented with a view to showing abstraction at its very idlest, the abstraction is not after all entirely senseless. It really does represent a step in sound reasoning.

Before going on to consider mathematical abstractions, let us ask ourselves how an *abstraction,* meaning that which an abstract noun denotes, is to be defined. It would be no proper definition of it to say that it is that which an abstract noun denotes. That would not be an analysis but a device for eluding analysis, quite similar to the old leach's offering *dormitive virtue* as an explanation of opium's putting people asleep. An abstraction is something denoted by a noun substantive, something having a name; and therefore, whether it be a reality or whether it be a figment, it belongs to the category of *substance,* and is in proper philosophical terminology to be called a *substance,* or thing. Now then let us ask whether it be a real substance or a fictitious and false substance. Of course, it may chance to be false. There is no magic in the operation of abstraction which should cause it to produce only truth whether its premiss is true or not. That, then, is not in question. But the question is whether an abstraction *can* be real. For the moment, I will abstain from giving a positive answer to this question; but will content myself with pointing out that upon pragmatistic principles an abstraction may be, and normally will be, *real.* For according to the pragmatistic maxim this must depend upon whether all the practical consequences of it are true. Now the only practical consequences there are or can be are embodied in the statement that what is said about it is *true.* On pragmatistic principles *reality* can mean nothing except the *truth* of statements in which the real thing is asserted.[7] To say that opium has a dormitive virtue means nothing and can have no practical consequences except what are involved in the statement that there is some circumstance connected with opium that explains its putting people to sleep. If there truly be such a circumstance, that is all that it can possibly mean,—according to the pragmatist maxim,—to say that opium really has a dormitive virtue. Indeed, nobody but a metaphysician would dream of denying that opium *really* has a dormitive virtue. Now it certainly cannot *really* have that which is pure figment. Without, then, coming to a positive decision as yet, since the truth of pragmatism is in question, we shall if we incline to believe there is something in pragmatism also incline to believe that an abstraction may be a real substance. At the same time nobody for many centuries,—unless it was some crank,—could possibly believe that an abstraction was an ordinary primary substance. You couldn't load a pistol with dormitive virtue and shoot it into a breakfast roll. Though it is in opium, it is wholly and completely in every piece of opium in Smyrna, as well as in every piece in every joint in the Chinatown of San Francisco. It has not that kind of existence which makes things *hic et nunc.* What kind of being has it? What does its reality consist in? Why it consists in something being true of something else that has a more primary mode of substantiality. Here we have, I believe, the materials for a good definition of an abstraction.

An abstraction is a substance whose being consists in the truth of some proposition concerning a more primary substance.

By a primary substance I mean a substance whose being is independent of what may be true of anything else. Whether there is any primary substance in this sense or not we may leave the metaphysicians to wrangle about.

By a *more* primary substance I mean one whose being does not depend upon all that the being of the less primary substance [does] but only upon a part thereof.

Now then armed with this definition I will take a shot at the abstractions of geometry [and] endeavor to bring down one or two of them.

We may define or describe a point as a place that has no parts. It is a familiar conception to mathematicians that space may be regarded as consisting of points. We will find that it is not true; but it will do for a rough statement.

We may define a *particle* as a portion of matter which can be, and at every instant of time is, situated in a point.

According to a very familiar conception of matter,—whether it be true or not does not concern us,—every particle is supposed to exist in such a sense that all other matter might be annihilated without this particle ceasing to exist. Supposing that to be the case the particle is, so far at least as matter is concerned, a primary substance.

Now let us imagine that a particle moves. That is, at one instant it is in one point[,] at another in another. That may pass as a concrete description of what happens but it is very inadequate. For according to this, the particle might be now here now there without continuity of motion.

But the geometer says[:] The place which a moving particle occupies on the whole in the course of time is a *line*. That may be taken as the definition of a line. And a portion of matter which at any one instant is situated in a line may be called a *filament*.

But somebody here objects. Hold, he says. This will not do. It was agreed that all matter is particles. What then is this filament? Suppose the objector is told that the filament is composed of particles. But the objector is not satisfied. What do you mean by *being composed*? Is the filament a particle? No. Well then it is not matter, for matter is particles. But my dear sir the filament is particles. Then it is not one but many and this single filament you speak of is a fiction. All there is is particles. But my dear sir, do you not understand that although all there is is particles yet there really is a filament because to say that the filament exists is simply to say that particles exist. Its mode of being is such that it consists in there being a particle in every point that a moving particle might occupy.

Thus you see that if the particles be conceived as primary substances the filaments are abstractions, that is, they are substances the being of any one of which consists in something being true of some more primary substance or substances none of them identical with this filament.

A *film* or that portion of matter that in any one instant occupies a surface will be still more abstract. For a film will be related to a filament just as a filament is related to a particle.

And a solid body will be a still more extreme case of abstraction.

Atoms are supposed to have existences independent of one another. But in that case according to our definition of an *abstraction,* a collection of atoms, such as are all the things we see and handle are *abstractions.* They are just as much abstractions as that celebrated jack-knife that got a new blade and then a new handle and was finally confronted with a resurrected incarnation of its former self.[8]

There is no denying this I believe, and therefore I do not think that we need have any further scruple in admitting that abstractions may be real,—indeed, a good deal less open to suspicion of fiction than are the primary substances. So the pragmatistic decision turned out correct in this instance, though it seemed a little risky at first.

That a *collection* is a species of abstraction becomes evident as soon as one defines the term *collection.* A *collection* is a substance whose existence consists in the existence of certain other things called its *members.*

An abstraction being a substance whose existence consists in something being true of something else, when this truth is a mere truth of existence the abstraction becomes a collection. When we reflect upon the enormous rôle enacted in mathematics by the conception of *collection* in all its varieties, we can guess that were there no other kinds of abstractions in the science (instead of the hosts of them that there are) still the logical operation of abstraction would be a matter of prime importance in the analysis of the logic of mathematics.

I have so much more to say than I have of time to say it in that all my statements have to be left in the rough and I know I must produce an impression of vagueness and haziness of thought that would disappear upon close examination. I shall be obliged to presume that after leaving the lecture room you will do some close thinking on your own accounts.

I have no time to speak further of the interesting and important subject of the reasoning of mathematics. Nor can I discuss Dedekind's suggestion that pure mathematics is a branch of logic. It would I think be nearer the truth (although not strictly true) to say that necessary reasoning is not one of the topics of logical discussion. I am satisfied that all necessary reasoning is of the nature of mathematical reasoning. It is always diagrammatic in a broad sense although the wordy and loose deductions of the philosophers may make use rather of auditory diagrams, if I may be allowed the expression, than with visual ones. All necessary reasoning is reasoning from pure hypothesis, in this sense, that if the premiss has any truth for the real world that is an accident totally irrelevant to the relation of the conclusion to the premiss; while in the kinds of reasoning that are more peculiarly topics of logical discussion it has all the relevancy in the world.

But I must hurry on to the consideration of the different kinds of mathematics, a subject of which the slightest sketch keeping close to what is wanted for the study of pragmatism ought in itself to occupy three good lectures at least, and would be much more interesting.

The different branches of mathematics are distinguished by the different kinds of fundamental hypotheses of which they are the developments.

The simplest conceivable hypothesis is that of a universe in which there is but one thing say A and nothing else whatever of any kind. The corresponding mathematics consists of a single self-evident proposition (that is it becomes evident by logical analysis simply) as follows: Nothing whatever can be predicated of A and it is absolutely indistinguishable from *blank nothingness*. For if anything were true of A, A would have some character or quality which character or quality would be something in the universe over and above A.

The next simplest mathematics seems to be that which I entitle *dichotomic mathematics*. The hypothesis is that there are two things distinguished from one another. We might call them B and M these being the initials of *bonum* and *malum*. Then the problem of this mathematics will be to determine in regard to anything unrecognized, say *x,* whether it is identical with B or identical with M. It would be a mere difference of phraseology to say that there are countless things in the universe, *x, y, z,* etc. each of which has one or other of the two values B and M. The first form of statement is preferable for reasons I cannot stop to explain.

The Boolian algebra of logic is a mere application of this kind of pure mathematics. It is a form of mathematics rather poverty-stricken as to ideas. Nevertheless, it has some features which we shall find have a certain bearing upon the foundations of pragmatism.

In the first place, although the universe consists of only two primary substances, yet there will *ipso facto* be quite a wealth of abstractions. For in the first place there will be the universe of which M and B are the two parts. Then there will be three prominent *relations*. Namely 1st the relation that M has to B and that M has to nothing else, and that nothing but M has to anything. 2nd there will be the converse relation that B has to M and to nothing else and that nothing but B has to anything, and 3rd the relation that B has to M and to nothing else and that M has to B and to nothing else. Without counting the absurd relation that nothing has to anything. That third relation is the self-converse relation of *otherness*.

Those four relations are *dyadic relations.* That is, considered as abstractions their existence consists in something being true of two primary substances. Thus to say that M is in the relation of otherness to B is to say that M is other than B which is a fact about the two primary substances M and B.

But there are also *triadic relations.* It is true that owing to there being but two primary substances, there is no triadic relation between three different primary substances. But there can be a relation between three different dyadic relations. There are dyadic relations between dyadic relations. Thus the relation of M to B is the *converse* of the relation of B to M and this relation of *converseness* is a dyadic relation between relations.

As an example of a triadic relation between relations take the relation between the 1st, 2nd and 3rd relations between M and B. That is the relation between first the relation of M only to B only, of B only to M only, and of otherness of M to B and of B to M. This triadic relation is a case of the general triadic relation of *aggregation.* To say that Z stands in the relation of aggregation to X and Y is to say that Z is true wherever X is true and where Y is true and that either X or Y is true wherever Z is true.

Another important kind of triadic relation between dyadic relations is where R is in the relation of relative product of P into Q where P, Q, R are dyadic relations. This means that if anything A is in the relation R to anything C there is something B such that A is in the relation P to B while B is in the relation Q to C and conversely if A is not in the relation R to C then taking anything whatever β either A is not P to β or else β is not Q to C.

Applying this idea of a relative product we get the conception of *identity* or the relation which M has to M and to nothing else and that B has to B and to nothing else.

I have only noticed a few of the most interesting of these abstractions. But I have not mentioned the most interesting of all the dyadic relations, that of *inclusion,* the great importance of which, now generally recognized, was first pointed out and demonstrated by me in 1870.

It is the relation that M has to M and to B and that B has to B but that B does not have to M. It is the connecting link between the general ideas of logical dependence and the idea of the sequence of quantity.[9]

All these ideas may be said to have virtually existed in the form of the Boolian algebra originally given by Boole. But in 1870 I greatly enlarged and I may say revolutionized the subject by the virtual introduction of an entire new kind of *abstractions.*

10

Prolegomena to an Apology for Pragmaticism

[Peirce 1906] Peirce opens this defense of his "pragmaticism" (so named to distinguish it from the pragmatisms of James and others)[1] with an imaginary debate over the diagrammatic nature of exact thought. He insists very literally on the importance of *experimentation* upon diagrams, and is undeterred by an apparent disanalogy between the diagrammatic experiments of the mathematician and the physical experiments of the chemist. In both cases the true object of study is a structure—mathematical or chemical—and the investigation proceeds by manipulating an object—a diagram, a chemical sample—which has the structure under investigation. A great deal of Peirce's philosophy of mathematics is packed into these few pages; it has been partially unpacked in the introduction (xxxv–xli).

Like any experimental method, the mathematician's leaves room for error; mathematical truths, though necessary, are not known with certainty. But they are at least approximately apodictic because fresh experimental subjects "can be multiplied *ad libitum* at no more cost than a summons before the imagination." Peirce's "rapid sketch" of a proof of the approximate apodicticity of mathematics begins with a brisk review of his primary semiotic triad (icon, index, symbol). Symbols, he says, can only express what is already known, and indices can give us no insight into their objects. Icons, on the other hand, are uniquely well adapted to the aims and requirements of diagrammatic reasoning. They do not require an actual object, as indices do. Their objects need only be logically possible, an ontological standard to which the mathematician's hypothetical states of things can rise. Indeed a diagram is "precisely an Icon of intelligible relations" of the sort the mathematician is interested in. Icons are therefore "specially requisite for reasoning" and most especially for necessary reasoning, in which "the conclusion follows from the form of the relation set forth in the premiss." This does not immunize us from error: "what must be is not to be learned by simple inspection of anything." But we can fall back on the facility with which we can replicate our iconic experiments *ad libitum;* this makes mathematical reasoning as error-free as an experimental inquiry can hope to be.

Come on, my reader, and let us construct a diagram to illustrate the general course of thought; I mean a System of diagrammatization by means of which any course of thought can be represented with exactitude.

"But why do that, when the thought itself is present to us?" Such, substantially, has been the interrogative objection raised by more than one or two superior intelligences, among whom I single out an eminent and glorious General.

Recluse that I am, I was not ready with the counter-question, which should have run, "General, you make use of maps during a campaign, I believe. But why should you do so, when the country they represent is right there?" Thereupon, had he replied that he found details in the maps that were so far from being "right there," that they were within the enemy's lines, I ought to have pressed the question, "Am I right, then, in understanding that, if you were thoroughly and perfectly familiar with the country, as, for example, if it lay just about the scenes of your childhood, no map of it would then be of the smallest use to you in laying out your detailed plans?" To that he could only have rejoined, "No, I do not say that, since I might probably desire the maps to stick pins into, so as to mark each anticipated day's change in the situations of the two armies." To that again, my sur-rejoinder should have been, "Well, General, that precisely corresponds to the advantage of a diagram of the course of a discussion. Indeed, just there, where you have so clearly pointed it out, lies the advantage of diagrams in general. Namely, if I may try to state the matter after you, one can make exact experiments upon uniform diagrams; and when one does so, one must keep a bright lookout for unintended and unexpected changes thereby brought about in the relations of different significant parts of the diagram to one another. Such operations upon diagrams, whether external or imaginary, take the place of the experiments upon real things that one performs in chemical and physical research. Chemists have ere now, I need not say, described experimentation as the putting of questions to Nature. Just so, experiments upon diagrams are questions put to the Nature of the relations concerned." The General would here, may be, have suggested, (if I may emulate illustrious warriors in reviewing my encounters in afterthought,) that there is a good deal of difference between experiments like the chemist's, which are trials made upon the very substance whose behavior is in question, and experiments made upon diagrams, these latter having no physical connection with the things they represent. The proper response to that, and the only proper one, making a point that a novice in logic would be apt to miss, would be this: "You are entirely right in saying that the chemist experiments upon the very object of investigation, albeit, after the experiment is made, the particular sample he operated upon could very well be thrown away, as having no further interest. For it was not the particular sample that the chemist was investigating; it was the molecular *structure.* Now he was long ago in possession of overwhelming proof that all samples of the same molecular structure react chemically in exactly the same way; so that one sample is all one with another. But the object of the chemist's research, that upon which he experiments, and to which the question he puts to Nature relates, is the Molecular Structure, which in all his samples has as complete an identity as it is in the nature of Molecular Structure ever to possess. Accordingly, he does, as you say, experiment upon the Very Object under investigation. But if you stop a moment to consider it, you will acknowledge, I think, that you slipped in implying that it is otherwise with experiments made upon diagrams. For what is there the Object of Investigation? It is the *form of a relation.* Now this Form of Relation is the very form of the relation between the two corresponding parts of the diagram. For example, let f_1 and f_2 be the two distances of the two foci of a lens from the lens. Then,

$$\frac{1}{f_1} + \frac{1}{f_2} = \frac{1}{f_0}.$$

This equation is a diagram of the form of the relation between the two focal distances and the principal focal distance; and the conventions of algebra (and all diagrams, nay all pictures, depend upon conventions) in conjunction with the writing of the equation, establish a relation between the very *letters* f_1, f_2, [f_0], regardless of their significance, the form of which relation is the *Very Same* as the form of the relation between the three focal distances that these letters denote. This is a truth quite beyond dispute. Thus, this algebraic Diagram presents to our observation the very, identical object of mathematical research, that is, the Form of the harmonic mean, which the equation aids one to study. (But do not let me be understood as saying that a Form possesses, itself, Identity in the strict sense; that is, what the logicians, translating ἀριθμῷ, call 'numerical identity.')"

Not only is it true that by experimentation upon some diagram an experimental proof can be obtained of every necessary conclusion from any given Copulate of Premises, but, what is more, no "necessary" conclusion is any more apodictic than inductive reasoning becomes from the moment when experimentation can be multiplied *ad libitum* at no more cost than a summons before the imagination. I might furnish a regular proof of this, and am dissuaded from doing so now and here only by the exigency of space, the ineluctable length of the requisite explanations, and particularly by the present disposition of logicians to accept as sufficient F. A. Lange's persuasive and brilliant, albeit defective and in parts even erroneous, apology for it.[2] Under these circumstances, I will content myself with a rapid sketch of my proof. First, an analysis of the essence of a sign, (stretching that word to its widest limits, as *anything which, being determined by an object, determines an interpretation to determination, through it, by the same object,*) leads to a proof that every sign is determined by its object, either first, by partaking in the characters of the object, when I call the sign an *Icon;* secondly, by being really and in its individual existence connected with the individual object, when I call the sign an *Index;* thirdly, by more or less approximate certainty that it will be interpreted as denoting the object, in consequence of a habit (which term I use as including a natural disposition), when I call the sign a *Symbol.*[*] I next examine into the different efficiencies and inefficiencies of these three kinds of signs in aiding the ascertainment of truth. A Symbol incorporates a habit, and is indispensable to the application of any *intellectual* habit, *at least.* Moreover, Symbols afford the means of thinking about thoughts in ways in which we could not otherwise think of them. They enable us, for example, to create Abstractions, without which we should lack a great engine of discovery. These enable us to count, they teach us that collections are individuals (individual = individual object), and in many respects they are the very warp of reason. But since symbols rest exclusively on habits already definitely formed but not furnishing any observation even of themselves, and since knowledge is habit, they do not enable us to add to our knowledge even so much as a necessary consequent, unless by means of a definite preformed habit. Indices, on the other hand, furnish positive assurance of the reality and the nearness of their Objects. But with the assurance there goes no insight into the nature of those Objects. The same Perceptible may, however, function doubly as a Sign. That footprint that Robinson Crusoe found in the sand, and which has been stamped in the granite of fame, was an Index to him that some creature was on his island, and at the same time, as a Symbol, called up the idea of a man. Each *Icon* partakes of some more or less overt character of its Object. They, one and all, partake of the most overt character of all lies and deceptions,—their Overtness. Yet they have more to do with the living character of truth than have either Symbols or Indices. The Icon does not stand unequivocally for

---

*. In the original publication of this division, in 1867, the term "representamen" was employed in the sense of a sign in general, while "sign" was taken as a synonym of *index,* and an *Icon* was termed a "likeness."

this or that existing thing, as the Index does. Its Object may be a pure fiction, as to its existence. Much less is its Object necessarily a thing of a sort habitually met with. But there is one assurance that the Icon does afford in the highest degree. Namely, that which is displayed before the mind's gaze,— the Form of the Icon, which is also its object,—must be *logically possible.* This division of Signs is only one of ten different divisions of Signs which I have found it necessary more especially to study.[3] I do not say that they are all satisfactorily definite in my mind. They seem to be all trichotomies, which form an attribute to the essentially triadic nature of a Sign. I mean because three things are concerned in the functioning of a Sign; the Sign itself, its Object, and its Interpretant. I cannot discuss all these divisions in this article; and it can well be believed that the whole nature of reasoning cannot be fully exposed from the consideration of one point of view among ten. That which we can learn from this division is of what sort a Sign must be to represent the sort of Object that reasoning is concerned with. Now reasoning has to make its conclusion manifest. Therefore, it must be chiefly concerned with forms, which are the chief objects of rational insight. Accordingly, Icons are specially requisite for reasoning. A Diagram is mainly an Icon, and an Icon of intelligible relations. It is true that what must be is not to be learned by simple inspection of anything. But when we talk of deductive reasoning being necessary, we do not mean, of course, that it is infallible. But precisely what we do mean is that the conclusion follows from the form of the relations set forth in the premiss. Now since a diagram, though it will ordinarily have Symbolide Features, as well as features approaching the nature of Indices, is nevertheless in the main an Icon of the forms of relations in the constitution of its Object, the appropriateness of it for the representation of necessary inference is easily seen.

11

['Collection' in The Century Dictionary*]*

[Whitney 1889; Ms 1597; Peirce 1888(?)–1914(?); Whitney 1909] The original definition of 'collection' in *The Century Dictionary* was written, not by
Peirce, but by an anonymous contributor; it is reproduced here for reference.
Peirce's critical notes on that definition, written in his own copy of the dictionary, date from some time around the turn of the century; they are followed here by his definitions of two additional senses of 'collection' for the
1909 *Supplement* to the dictionary.

The notes open with a definition:

> A *collection* (or plural) is an individual object whose existence consists in
> the existence of whatever individuals there may exist having one character,
> these being called the *members* of the collection.

This is a dry run for, but differs importantly from, the definition of the second new sense Peirce contributed to the *Supplement:*

> A plural object; an individual object whose existence consists in the exist-
> ence of whatever individuals may have been mentally connected and
> regarded as parts of it.

The latter is more mentalistic than the former; the approach Peirce takes in
the notes, which makes collections ontologically dependent on their defining
characters, is more typical of the selections that follow this one. Since the
Supplement postdates those, the second new sense may betoken a major doctrinal change, or else may just be a "sop to Cerberus" forced upon Peirce by
the space constraints of a dictionary entry.

In the notes Peirce gives extensional identity criteria for collections,
thereby differentiating them from such intensional entities as characters.
(Note that for Peirce a *set* is an ordered collection, by marked contrast with
what has since become standard usage.) In his notes Peirce recognizes the
empty collection *as* a collection; he also distinguishes a collection with one
member from that member, a collection from its defining attribute, and a collection of collections from the union of its members. All of these ideas are
now basic to the conceptual equipment of set theory.

A collection in the first new sense is its members *regarded as* an individual object, whereas in the second sense a collection *is* an individual object. The first new sense is very close to one of Cantor's definitions of set ("[a] multiplicity which can be thought of as one" (Cantor 1883, 916)); Peirce may have separated the senses in order to distinguish his view from Cantor's. According to both the annotations and the entry, a collection's existence consists in that of its members. Yet though sameness of members is both necessary and sufficient for sameness of collections, Peirce adds in both the notes and the entry that every collection has a defining character (an essence), which is intensional. In the dictionary entry Peirce tries to sort out the relationship between existence and essence for collections.[1] Logicians think of their collections as comprising distinct individuals, whose properties may change without changing the identity of the collections. Mathematicians, on the other hand, deal with collections of hypothetical objects whose identities are more sensitive to changes of property. As a result, essence and existence are not as distinct for a mathematician's collections as they are for a logician's. The selections that follow take up these ontological problems at greater length; it is left to the reader to determine whether this entry, which as just noted is later than them all, shows that Peirce's metaphysics of collections had reached a reflective equilibrium or was still on the way to one.

1. 'COLLECTION' FROM THE CENTURY DICTIONARY

2. An assemblage or gathering of objects; a number of things collected, gathered, or brought together; a number of objects considered as constituting one whole of which the single objects are parts: as, a *collection* of pictures; a *collection* of essays; a *collection* of minerals.

2. PEIRCE'S CRITICAL ANNOTATIONS

collection 2. Not a very successful definition. A *collection* (or plural) is an individual object whose existence consists in the existence of whatever individuals there may exist having one character, these being called the *members* of the collection. It is commonly limited to the case in which more than one individual exist[s] having that character. It is, however, desirable to consider a collection which happens to contain one individual as different from that individual itself. Moreover, *nothing* is that sole collection of which no member exists; and therefore this collection, though named, has no *exist-ence,* in one sense. A collection differs from a *set* in not being altered by any transposition of members. It differs from a *character* in consisting in the existence of the members, so that two different collections must have differ-ent members (i.e. there must be a member belonging to one that does not belong to the other,) while two characters may belong to precisely the same subjects.

There may be a collection of collections which will be entirely different from the aggregate collection of the members of those collections.

Thus, ten companies of soldiers make one thing and the soldiers that make up those companies make another thing.

Properly, the characters of a collection ought to depend upon the com-mon characters of all the members. But in ordinary language we generally make it depend upon the characters of the members that are most remark-able. Thus, a poetical people would mean a people of whom perhaps one percent of the individuals were poetical.

3. PEIRCE'S DEFINITION

8. In *logic,* many independent or discrete objects regarded as a single object composed of these objects. In this sense 'many' is to be taken as including the case of a single object regarded as being composed of itself alone. The modern logicomathematical science of multitude (often called the *theory of cardinal numbers*) relates to the magnitudes of collections.[2]

9. A plural object; an individual object whose existence consists in the existence of whatever individuals may have been mentally connected and regarded as parts of it. Different logicians and mathematicians have different objects in mind in speaking of a 'collection,' without always recognizing that they are at cross-purposes. Most logicians are in the habit of thinking of objects as they would be if they were real, so that each is assumed to be in itself definitely distinguished from every other. With them, the identity of a collection lies in the identity of its individual members; so that whatever metamorphoses the different individuals might undergo, as long as their identities were conserved, that of the collection would remain. But if an indi-vidual member is destroyed or a new one created, a different collection is produced, though the definition of the *class* (which is a collection recognized as consisting of whatever existent objects possess a certain common charac-ter) may be unchanged. Most writers on pure mathematics, on the other

hand, are in the habit of studying objects that are purely hypothetical, without any consideration of whether any such objects exist. These objects, being mere creatures of thought, possess only such individuality as is determinately predicated of them. They are what the logicians term 'indesignate individuals,' a name which fails to recognize the extrinsic, superimposed character of their individuality. A mathematician, for example, will think of a collection consisting of a dot, of a dot expressly supposed to be other than the former, and of a dot expressly supposed to be neither of the others. These dots, being mere creatures of thought, are entirely alike as long as they are not thought as unlike. But being expressly supposed to be each other than either of two, so they necessarily are in their hypothetical being. Thus the mathematician's collection, being a mere creature of thought, changes its identity as soon as it is altered at all, unless it be expressly supposed to remain the same collection. The logician's collection is also created by thought, but it is thought to exist in the real existence of its individual members. Thus the logician's collection has a derived existence distinct from its essence, which latter lies in the intention of the act of thought which severs the universe into two portions, the one to form the inside and the other the outside of the collection. Accordingly, a logician's collection may contain but a single member with which the collection is identical in existence, although its essence refers also to everything excluded. So, too, if to the question, 'What is in this box?' the answer be, 'Nothing', this word, as a reply to that question, signifies the essence of a collection, namely of the one sole logical collection which has no existence.

12

[On Collections and Substantive Possibility]

[Peirce 1903g] Peirce delivered two major series of lectures in 1903: the Harvard Lectures on Pragmatism, and a series at the Lowell Institute on "Some Topics of Logic bearing on Questions now Vexed." This rejected draft of the third Lowell Lecture begins with a polished review of Peirce's definition of mathematics, and of the classification of mathematical specialties by the number of "alternatives" they deal with. Peirce then asserts, without explanation, that the whole numbers are the subject matter of the next branch of mathematics after the simplest mathematics, that of two alternatives. The explanation may well be found at the opening of the third Harvard Lecture on Pragmatism, where Peirce says that the whole science of multitude (which begins with the natural numbers) involves "a mere complication of Category the Third" (Peirce 1903a, E2.160).

Peirce accepts, with terminological qualifications, Cantor's distinction between cardinal and ordinal numbers. The qualification is that Peirce restricts 'cardinal number' to the *words* (numerals, etc.) used in counting collections of objects. What Cantor calls "cardinal number" he calls "multitude": this is a characteristic of a collection, its "degree of maniness." He does not quarrel with Cantor's doctrine of ordinal numbers, though he gives it the formalistic interpretation Cantor himself put forth in his early publications on the subject, but later abandoned in favor of a more realistic view (Dauben 1990, pp. 81–82, 98–99).

The theory of ordinals, Peirce says, belongs to pure mathematics, but that of cardinals does not. His argument for this claim (Peirce 1903g, N3.347–350) turns on notational details of his existential graphs, and is omitted here. (He gives an alternative argument, without recourse to the graphs, in selection 13.) The idea is that the graph that represents the relation of posteriority, which underlies the sequence of ordinal numbers, is an elaboration of the graph that represents the relation of inclusion, which "enters into the very definition of necessary reasoning" (349). (For more on inclusion, and its connection with number, see note 9 to selection 9, p. 246.) On the other hand, the doctrine of multitude (cardinal number) belongs not to pure mathematics but rather to logic: "Multitudes are characters of collections; and the idea of a collection is essentially a logical conception."[1]

Peirce's definition of 'collection' is prefaced by a lengthy disquisition on what he calls the doctrine of "substantive possibility." This is that part of logic which studies qualities and relations, which "are possibilities of a peculiar kind." We can think of this doctrine as a philosophically informed branch of modal logic, though Peirce only hints here at its technical development. At first blush it seems that he is mainly concerned here with the metaphysical underpinnings of the doctrine, but he repeatedly insists that he is doing logic and not metaphysics. The justification for the "metaphysics" is its indispensability to a workable logical theory:

> You can entertain whatever opinion seems good to you as to the real nature of qualities and as to the genesis of ideas in the mind. I have in this course quite nothing at all to say to all that. I simply say that you must use this form of thought, whether you regard it as corresponding to facts literally, metaphorically, symbolically, or however you may prefer. But you must use the form of thought or your threads will be inextricably entangled.

The quality *red* as a substantive possibility is not the redness of a particular apple, but rather the logical possibility which the apple very partially realizes. The quality may be said to "exist" when it has a "replica" of this sort; but the being of the quality is independent of its replication, and "consists in the fact that a thing *might be*" (or as he puts later, "consists in such logical possibility as there may be that a definite predicate should be true of a single subject"). Qualities as such, however, are prior to existence: they are, in the vocabulary of Peirce's categories, Firsts rather than Seconds. Since only Seconds, which can causally interact, can have distinct identities, qualities do not have distinct identities; red, for instance, takes in many shades of red, which shade into one another without any sharp boundaries. (A quality is thus a continuum in the sense that Peirce strives to clarify in selections 18–29.) Moreover, though we are directly aware of a relatively small number of qualities, there is no infinite multitude large enough to express the size of the universe of qualities. (Here again, Peirce's conception of a continuum as a "supermultitudinous collection" is in the background.) He asserts without argument that higher order qualities—qualities of qualities, etc.—are less multitudinous than lower order ones. Having stated his official definition of 'quality' in terms of logical possibility, Peirce takes up the ontology of qualities. They are held to be *entia rationis* because they consist in the meaningfulness (as opposed to the truth) of propositions. A cryptic definition of essence allows Peirce to claim that qualities have essence but not existence (which requires "brute compulsion").

His definition of 'collection' makes use of this metaphysical (or, as Peirce would have it, logical) machinery. Collections, like qualities, have being and *may* have existence; to have being they must have essence, and to have existence they must have members (just as qualities, to exist, must have replicas). Peirce's first explanation of the being of collections is rather idealistic: it consists in the possibility of their members being thought together. But this is then held to be equivalent to the members sharing a defining quality, which is the essence of the collection. The empty collection is found to have essence but not existence. But the identity criteria for collections are clouded by a new complication: what happens when a collection's members change but its essence does not? This will be a major focus of selection 13.

Ladies and Gentlemen:

Mathematics is the science which draws necessary conclusions. Such was the definition first given by my father, Benjamin Peirce, in 1870. At that day the new mathematics was in its early infancy and the novelty of this definition was disconcerting even to the most advanced mathematicians; but today no competent man would adopt a definition decidedly opposed to that. The only fault I should find with it is that if we conceive a science, not as a body of ascertained truth, but, as the living business which a group of investigators are engaged upon, which I think is the only sense which gives a natural classification of sciences, then we must include under mathematics everything that is an indispensible part of the mathematician's business; and therefore we must include the *formulation* of his hypotheses as well as the tracing out of their consequences. Certainly, into that work of formulation the mathematicians put an immense deal of intellectual power and energy.

Moreover, the hypotheses of the mathematician are of a peculiar nature. The mathematician does not in the least concern himself about their truth. They are often designed to represent *approximately* some state of things which he has some reason to believe is realized; but he does not regard it as his business to find out whether this be true or not; and he generally knows very well that his hypothesis only approximates to a representation of that state of things. The substance of the mathematician's hypothesis is therefore a creature of his imagination. Yet nothing can be more unlike a poet's creation. The reason is that the poet is interested in his images solely on account of their own beauty or interest as images, while the mathematician is interested in his hypotheses solely on account of the ways in which necessary inferences can be drawn from them. He consequently makes them perfectly definite in all those respects which could affect the ways in which parts of them could or could not be taken together so as to lead to necessary consequences. If he leaves the hypotheses determinate in any other respects, they are hypotheses of *applied* mathematics. The pure mathematician generalizes his hypotheses so as to make them applicable to all conceivable states of things in which precisely analogous conclusions could be drawn. In view of this I would define Pure Mathematics as the science of pure hypotheses perfectly definite in all respects which can create or destroy forms of necessary consequences from them and entirely indeterminate in other respects.

I am confident that this definition will be accepted by mathematicians as, at least, substantially accurate. . . .

A mathematical reasoning may be defined as a reasoning in which the following of the conclusion does not depend on whether the premises represent experience, or represent the state of the real universe, or upon what universe it may be that they apply to. This erects, as we shall see, a definite party-wall between the reasoning of mathematics and much of the reasoning of all the positive sciences, including philosophy. But, of course, all the other sciences have recourse to the mathematician very frequently, and none so constantly as logic. There is no science more infested with a vermin of ignorant pretenders than logic; and there is one simple question by which they can commonly be detected. Ask your pretended logician whether there are any necessary reasonings of an essentially different character from mathematical reasonings. If he says no, you may hope he knows something about logic; but if he says "yes," he is contradicting a well-established truth universally admitted by sound logicians. If you ask for a sample, it will be found to be a very simple mathematical reasoning *blurred* by being confusedly apprehended. For a necessary reasoning is one which *would* follow under all circumstances, whether you are talking of the real world or the world of the 'Arabian Nights' or what. And that precisely defines mathematical reasoning. It is true that a *distinctively* mathematical reasoning is one that is so intricate that we need some kind of a diagram to follow it out. But something of the nature of a diagram, be it only an imaginary skeleton proposition, or even a mere noun with the ideas of its application and signification is needed in all necessary reasoning. Indeed one may say that something of this kind is needed in all reasoning whatsoever, although in induction it is the real experiences that serve as diagrams.

One of the most striking characters of pure mathematics,—of course you will understand that I speak only of mathematics in its present condition, and only occasionally and with much diffidence speculate as to what the mathematics of the future may be,—but one of [the] characters of latter day pure mathematics is that all its departments are so intimately related that one cannot treat of any one as it should be treated without considering all the others. We see the same thing in several other advanced sciences. But so far as it is possible to break mathematics into departments, we observe that in each department there is a certain set of alternatives to which every question relates. Thus, in projective geometry, which is the whole geometry that is allied to perspective *without measurement,* namely, the geometry of planes, their intersections and envelopes, and the intersections and envelopes of intersections and envelopes, the question always is whether a figure lies in another figure or not, whether a point one way described lies on a point another way described, whether a point lies on a line or not, whether three lines coincide or not. Here there are two alternatives. In other departments, the alternatives are all the integer numbers; in still others, the alternatives are all the analytical numbers, etc. The set of alternatives to which a branch of mathematics constantly refers may be considered as a system of values; and in that sense, mathematics seems always to deal with quantity. It would seem that if any lines of demarcation are to be drawn between different mathematical theories they must be according to the number of alternatives in the set of alternatives to which it refers; but I am bound to say that this is a notion personal to myself and that I have my doubts as to its worth as the basis of a complete classification of mathematics. We may, however, accept it in so far as it shows that the simplest possible kind of mathematics will be that all whose questions relate to which one of a single set of two alternatives is to be admitted. Now in Existential Graphs, all questions relate to whether a graph is true or false; and we may conceive that every proposition has one or other of two values, the *infinite* value of being true, and the *zero* value of being false. We have, therefore, in Existential Graphs an exposition of the simplest possible form of mathematics. It is Applied Mathematics, because we have given definite logical significations to the graphs. But if we were to define the graphs solely by means of the five fundamental rules of their transformation, allowing them to mean whatever they might mean while preserving these rules, we should then see in them the Pure Mathematics of two values, the simplest of all possible mathematics.

Were we to follow out the same principle, we should divide all mathematics according to the number of alternatives in the set of alternatives to which it constantly refers and also to the number of different sets of alternatives to which it refers. Perhaps that would give as natural a classification of pure mathematical inquiries as any that could at this time be proposed. At any rate, we may so far safely trust to it, to conclude that the very first thing to be inquired into in order to comprehend the nature of mathematics, is the matter of *number.*

Certainly, of all mathematical ideas, next after the idea of two alternatives, the most ubiquitous is the idea of whole numbers. Dr. Georg Cantor is justly recognized as the author of two important doctrines, that of *Cardinal Numbers* and that of *Ordinal Numbers.* But I protest against his use of the term *Cardinal Number.* What he calls cardinal number is not number at all. A cardinal number is one of the vocables used primarily in the experiment called counting a collection, and used secondarily as an appellative of that collection. But what Cantor means by a cardinal number is the *zeroness, oneness, twoness, threeness,* etc.,—in short the *multitude* of the collection. I shall always use the word *multitude* to mean the *degree of maniness* of a collection. By *ordinal numbers* Cantor means certain symbols invented by him to denote the place of an object in a series in which each object has another next after it. The character of being in a definite place in such a series may be called the *posteriority* of the object.

Since I have alluded to Cantor, for whose work I have a profound admiration, I had better say that what I have to tell you about Multitude is not in any degree borrowed from him. My studies of the subject began before his, and were nearly completed before I was aware of his work, and it is my independent development substantially agreeing in its results with his of which I intend to give a rough sketch. And since I have recommended Dedekind's work, I will say that it amounts to a very able and original development of ideas which I had published six years previously. Schröder in the third volume of his logic shows how Dedekind's development might be made to conform more closely to my conceptions. That is interesting; but Dedekind's development has its own independent value. I even incline to think that it follows a comparatively better way. For I am not so much in love with my own system as the late Professor Schröder was. I may add that quite recently Mr. Whitehead and the Hon. Bertrand Russell have treated of the subject; but they seem merely to have put truths already known into a uselessly technical and pedantic form.[2]

The two doctrines of Cardinal and Ordinal Numbers, or of Multitude and Posteriority, though necessarily running parallel are curiously unlike one another. . . . The doctrine of ordinal numbers . . . is a theory of pure mathematics and, as matters stand today, is the most fundamental of all branches of pure mathematics after the mathematics of the pair of values which existential graphs illustrate. The doctrine of multitude is not pure mathematics. Pure mathematics can see nothing in multitudes but a linear series of objects, having a first member, each one being followed by a next, with a few other such formal characters. A multitude, as such, Pure Mathematics knows nothing of. Multitudes are characters of collections; and the idea of a collection is essentially a logical conception. How would you define a collection, in general, without using the idea itself in your definition? It is not easy. In order to explain the matter, it is necessary to begin with the conception of a quality. There is an essential part of the doctrine of Existential Graphs,—essential to it, I mean, as a logical, not as a mathematical doctrine,—and of such importance as quite to overshadow all the rest which I have been forced to pass over for lack of time. It treats of the general properties of qualities and relations. Without it there are most important inferences that cannot be drawn. I call it the doctrine of *substantive possibility,* because qualities and relations are possibilities of a peculiar kind. In a secondary sense a quality may be said to exist when it has as it were, a replica in an existing thing. But strictly speaking, a quality does not exist. For to *exist* is to be a subject of blind compulsion. A quality not only neither exerts nor suffers such force, but it cannot even be called an *idea* of the mind. For things possess their qualities just the same, whether anybody thinks so or not. The being of a quality consists in the fact that a thing *might be* such or such like. In saying this, I am not talking metaphysics nor epistemology. I am confining myself to logic. You can entertain whatever opinion seems good to you as to the real nature of qualities and as to the genesis of ideas in the mind. I have in this course quite nothing at all to say to all that. I simply say that you must use this form of thought, whether you regard it as corresponding to facts literally, metaphorically, symbolically, or however you may prefer. But you must use the form of thought or your threads will be inextricably entangled. For my part, when I think about logic, I dismiss irrelevant questions of metaphysics and psychology entirely from my mind. But that requires some training. Qualities, then, and Relations are pure possibilities; and as such they have no individual identity. Two qualities are more or less unlike. Identity belongs only to subjects of blind compulsion or force. There is no sense for example in asking how many shades of red there are,—unless you mean how many a man can distinguish,—which is a question of psychophysics, not logic. These substantive possibilities,—that is, qualities, relations, and the like,—are *prior* to existence, in the sense that non-existence is not a necessary proof of non-possibility, but non-possibility is a necessary proof of non-existence. For it is logically impossible that existence should exhaust pure possibilities of any kind. These truths are strictly deducible from the facts of phenomenology, or the analysis of the phenomenon; meaning by the *phenomenon* whatever is present to the mind in any kind of thought. . . .

The doctrine of substantive possibility is an extensive one. You will understand that I only mention bits of it.

The variety of qualities, as is easily proved, literally exceeds not only all number, but all multitude finite or infinite; and anybody who should assume the contrary would be liable to great errors of reasoning. But the qualities with which we are familiar are a small number. Certainly a figure *one* with only twenty or thirty zeros after it would denote [a] greater number; and these are naturally regarded by us as composed of a *very* small number of qualities which we do not analyze. Therefore for all those purposes for which [we] regard the qualities themselves, they may be considered as comparatively few.

Qualties are general respects in which existing things *might* agree or differ. They are as I have said, mere possibilities. But qualities have themselves *general* respects in which they agree or differ. Thus, musical notes differ in respect to duration, intensity, pitch, timbre, stress, expression, and some other respects. These qualities of qualities differ very much from qualities of existing things. Considering the qualities of any one class of qualities, we find them to be innumerable indeed but not in excess of all multitude; and a set [of] three or four or some such small number of them that are independent of one another will fully suffice to describe the rest. These respects, or qualities of qualities, themselves again have general respects in which they agree and differ. Thus duration, pitch, and intensity are serial, that is each can only vary along one line of variation, while timbre, stress, and expression are multiform. These modes of variation of respects correspond to the possible whole numbers.

After these explanations, you will be able to understand this definition of a *quality*

> A quality is anything whose being consists in such logical
> possibility as there may be that a definite predicate should
> be true of a single subject.

It is said to be actually *embodied in* or *possessed by* whatever there is of which that predicate is true.

But somebody may ask, Has a quality any being? I reply, Why of course it must have being because by the terms of its definition to say that it has being is at the very most, no more than to say that something is logically possible. Remember, we are not talking metaphysics; we are talking logic. A quality is an *ens rationis* of course. That is, it consists in a certain proposition's having a meaning. The term *Essence* means being such as the subject of the essence necessarily is. Quality then has *essence*. But it has no *existence,* because it neither exercises nor suffers brute compulsion.

I am now prepared to give you the definition of a collection; and remember that by a collection I do not mean that whose members are in any sense actually brought together, nor even that whose members are actually thought together; but I mean that whose members *might,* in logical possibility, be thought together. But to think things together is to think that something might be true of them all that was true of nothing else. But to do this amounts to thinking that they have a common quality. Therefore the definition is plain:

A *Collection* is anything whose being consists in the existence of whatever there may exist that has any one quality; and if such thing or things exist, the collection is a single thing whose existence consists in the existence of all those very things.

According to this definition a *collection* is an *ens rationis.* If its members are actually brought together like the atoms which compose my body, it is more than a mere *collection.* As *collection,* it is an *ens rationis,* but that reason or *ratio* that creates it may be among the realities of the universe. A *collection* has *essence* and may have *existence.* It has *essence* from all eternity, in the logical possibility that it should be described. It has *existence* from the moment that all its members exist. Thus, all men constitute a collection; and not a very small one. But in the carboniferous period in a certain sense that collection had no existence. By saying it was so 'in a certain sense' I mean if by *men* be meant the men that live at the moment. In this same sense, the existence of this collection is constantly changing; the same collection in essence is becoming a different collection in existence. There is a collection of men with grass green hair; but having only essence and not existence, it has no individual identity. It is the collection that we call *Nothing.* It must be counted among collections; but it differs from all the rest in having no existence. Of course, for ordinary purposes, this is the emptiest nonsense. Nevertheless, it is a matter that has to be put straight for logical purposes. I may remark that nonsense often repays logical study and by that study enables us to avoid fallacious reasoning about serious questions. Another such little point is the following. According to the definition, there must be a collection of luminaries of the day. But there happens to be only one luminary of the day; namely, the Sun. Here then is a collection having but one member. Is not that collection the sun itself? I reply, Certainly not.

For a collection is an *ens rationis*. Its being consists in the truth of some-thing. But the Sun is not an *ens rationis* and its being does not consist in the truth of any proposition. It consists in the act of brute force in which it reacts with everything in its neighborhood. So then the Sun is one thing and exists, and this collection containing only the sun is something different and exists, and there would be a collection embracing as its sole member this collection, and this too exist[s] and so on *ad infinitum*. This is true. Yet there is only one *existence;* for the existence of the collection *is* the existence of its sole mem-ber. Thus, that collection embracing the sun alone is different from the Sun but its existence is the same as the existence of the sun. In that sense, it is the same as the Sun.

In the next lecture I will show you what *multitude* is and what different grades of multitude there are; and then you will see how some of the hair-splitting of this lecture is, after all, very useful.

13

[The Ontology of Collections]

[Peirce 1903d] There is a good deal of overlap between selection 12 and this one, which is taken from the fifth of the Lowell Lectures. The substantial differences, even in the areas of overlap, show that Peirce's views were evolving as he wrote. A noteworthy difference between the two presentations is that Peirce's categories play a much more explicit role in this attempt to define 'collection'. The centerpiece of this excerpt from the lecture is another attack on the problem of existence and essence for collections. Peirce distinguishes two senses of 'collection' by coining a new term for each. A *gath* has Secondness and no Firstness (existence but no essence). Its being consists in the existence of its members: it persists exactly as long as its members do, whatever alterations those members might undergo. It is in effect what contemporary logicians would call the mereological sum of its members. A *sam,* on the other hand, has Firstness—its essence or defining quality—and *may* have Secondness if there are actual objects having that quality. If there are such objects, they compose a gath whose existence is at the same time the existence of the sam.

For example, consider the village of Zenith, Ohio, where no one goes to college and everyone was born in the town. Suppose that at noon only residents of Zenith are within the city limits. At 12:01 Jim and Jane, two native Californians with Ph.D.s in Philosophy, drive into town, passing back out at 12:03; no one else enters or leaves town all day. Let Q_c be the quality of being a Californian in Zenith, Q_p that of being a Ph.D. in Zenith, and let S_c and S_p be the corresponding sams. At noon there is no gath for either of these sams; between 12:01 and 12:03 there is one gath, composed of Jim and Jane, for both. That gath existed before 12:01 and exists after 12:03, because Jim and Jane did and do; but except for their brief sojourn in Zenith it is the gath of sams other than S_c and S_p. Unless there happen to be some Californians in town, S_c has being but not existence, and similarly for S_p. As this example illustrates, a sam may lack a gath. A gath, on the other hand, must have a sam—this latter principle is, Peirce claims, "what we ordinarily express by saying that whatever exists is possible." Anything that exists (including a mereological sum) has a defining quality which gives rise to a sam.

In the light of this distinction Peirce gives a modified analysis of singleton collections. The sun is identical with its gath because the existence of both consists in the sun's existence; it is not, however, identical with its sam because the latter is an *ens rationis* with an essence, and the former is not. Peirce does not give a thorough analysis of the empty collection: though there clearly can be no empty gath, it is not clear whether any non-existent sam should be distinguished as *the* empty sam.

The nomenclature of sams and gaths did not get a firm grip in Peirce's writings on collections, though some form of the distinction is clearly at work in his entry on collections for the *Century Dictionary Supplement.* Surely Peirce's theory of continuity, which in one of its major phases involves collections whose members are not all actual, was a driving force behind his search for a sufficiently subtle ontology of collections. The tone of this lecture is confidently assertive. But a few years later, in selection 26, Peirce will somewhat plaintively suggest that the concept of collection may be "indecomposable."

But I must hasten to the subject of numbers.

Whole numbers can on the one hand be studied in two ways which are surprisingly different from one another throughout. They can be studied as qualities of collections, making the members of one collection many and those of another few, which [is] called by the Germans with their usual incapacity for language the doctrine of *Cardinal Numbers;* but which ought to be called the doctrine of *Multitude.* Or, on the other hand, numbers may be considered simply as objects in a sequence, as *ordinal numbers.* The latter study is a branch of pure mathematics, because it makes no difference what kinds of objects they are that are in series, nor whether it is a series in time, in space, or in logic. The doctrine of multitude, on the other hand, is not pure mathematics. For the objects it studies, the multitudes, are in a linear series exactly as the doctrine of ordinal numbers supposes; and since the doctrine of ordinal numbers permits the members of the series to be objects of any kind, it follows that it permits them to be multitudes. Thus the doctrine of multitude is nothing but a special application of the doctrine of ordinal numbers. But the special objects of its series have a special character which permits them to be studied from a special point of view; and that point of view is a logical point of view. It is not the pure mathematical forms that we study in the doctrine of multitude. It is on the contrary a branch of logic which, like all logic, is directly dependent upon mathematics.

The first question we come upon in the study of multitude is very obviously a purely logical question; and there is nothing at all corresponding to it in the doctrine of ordinal numbers: it is the question what is multitude. Multitude is obviously a relative quality of *collections,* or plurals. Therefore the question becomes, What is a *collection*? That is obviously a most important question for logic; and it is about as difficult a one as could be found. In speaking of a collection, we do not mean that its members are physically or in any way existentially brought together. We mean by a collection merely a plural, whose objects are collected together by thought. The collection exists just as much as its members. Their existence is its existence. Yet in another point of view, it is a creation of thought. It is an *ens rationis.* An *ens rationis* may be defined as a subject whose being consists in a Secondness, or fact, concerning something else. Its being is thus of the nature of Thirdness, or thought. Any abstraction, such as Truth and Justice, is an *ens rationis.* That does not prevent Truth and Justice from being real powers in the world without any figure of speech. They are powers, just as much, and in the same way, as I am a power if I can open my window should the air seem to me stuffy. A collection is an abstraction, or is like an abstraction in being an *ens rationis.* But it is unlike an abstraction in that it *exists.* Truth and justice do not *exist,* although they are powers. I myself, properly speaking, do not exist. It is only a replica of me that exists, and I exist in that replica as the effect of my being as a law. A collection, however, exists, and this existence is derived from the existence of its members which may be pure Secondness. Our bodies are of course much more than so many collections of molecules; but as far as its *existence* is concerned, the existence of our bodies consists in the existence of the molecules. But the word *collection* and other words of the same general meaning have two different meanings with a very fine distinction between them. This makes a large part of the difficulty of defining a collection and the non-recognition of this distinction makes a serious stumbling block in the doctrine of cardinal numbers. In accordance with my views of the Ethics of Terminology, I am going to make two new words to distinguish these two meanings. The one I shall call a gath which is simply the word 'gather' with the last syllable dropped. The other I call a 'sam' which is the word 'same' with the last letter dropped. I also like this word because it is so much like the word *sum,* in the phrase *sum total.* It also recalls the German word *Samlung.*[1] A collection, in the sense of a *gath* is a subject which is a pure Secondness without Firstness, and whose only mode of being is whatever existence it may have; and this consists in the existence of certain other existents, or pure Seconds, called its *members.* Thus, the gath of human beings at this moment in Boston, consists in the existence of this man and that man. No matter how those men might be transformed, no matter if some of them were to leave Boston, that same gath exists, although it would cease to be the gath of the inhabitants of Boston. But were a single member of the collection to cease to exist, that gath would no longer exist. There would still be a gath of inhabitants of Boston but it would be a different gath. The description, 'All the inhabitants of Boston' describes a gath. But as time goes on it will describe a different gath. The description 'the inhabitants of Boston' is a proper name. It applies to but a single individual object, the whole of all the inhabitants of Boston. This whole is what I call a *Sam* [and it] is not exactly a gath; and it is important to get a distinct idea of the difference. Just as the molecules that compose a man's body are continually changing by the loss of some and the gain of others, while there remains the same man, so the population of Boston is ever changing, yet remains the same individual whole. I propose to say it is the same *sam.* But it does not remain the same gath. At each instant it is identical with a gath. Always there is a gath in the existence of which consists the existence of the sam of the inhabitants of Boston. Were the city to be devastated and not one inhabitant left, still, as long as it remained Boston, the 'sam, or sum total, of the inhabitants of Boston' would have a *being,* although it would under those circumstances have ceased to *exist.* It would continue to *be,* since the description would retain its meaning. The *essence* of the sam would remain, although its existence had departed. But as for the gath, since it has no other being than existence, and its existence consists in the existence of its members; and since under those circumstances no members would exist, the gath would altogether cease to be. It is important to have this distinction clearly in mind. I do not mean to say that [it] is usually important to hold this distinction clear in regard to any collection that we may happen to speak of; but I mean that for certain purposes it is indispensibly necessary. Whatever sam there may be to whose members, and to them alone, any sign applies, is called the *breadth* of the sign. This word *breadth,* originating with the Greek commentators of Aristotle, has passed into our vernacular. We speak of a man of *broad* culture. That means *culture* in many fields. *Breadth* of mind is the character of a mind that takes many things into account. If a man has *broad* and *deep* learning, the breadth consists in how many different subjects he is acquainted with, and the depth in how much he knows about whatever subject he is acquainted with. Now the *breadth* of a descriptive appellation has an *essence,* or Imputed Firstness, which is the signification, or *Depth,* of the appellation.[2] Take the word *phenix.* No such thing exists. One naturally says that the name has *no breadth*. That, however, is not strictly correct. We should say *its breadth is nothing*. That breadth is precisely what I mean by a *sam*. Therefore I define a *sam* as an *ens rationis* having two grades of being, its *essence,* which is the being of a definite quality imputed to the sam, and its *existence* which is the existence of whatever subject may exist that possesses that quality. A *gath,* on the other hand is a subject having only one mode of being which is the compound of the existence of subjects called the *members* of the *gath.*

You may remark that a *sam* is thus defined without any reference at all to a gath. I repeat the definition, so that you may observe this:

A *sam* is an *ens rationis* whose essence is the being of a definite quality (imputed to the *sam*) and whose existence is the existence of whatever subject there may be possessing that quality.

On the other hand, it is impossible to define a gath without reference to a sam. For when I say that a gath is a subject whose only mode of being is the compounded existence of definite individuals called its members, what is the meaning of this *compounded* existence. It is plain that the idea of a compound is a triadic idea. It implies that there is some sign, or something like a sign, which picks out and unites those members. Now the fact that they are all united in that compound is a quality belonging to them all and to nothing else. There is thus here a reference to a possible *sam* which does this. Thus, we might as well at once [have] defined a *gath* as a subject which has but one mode of being which is the existence of a *sam*. From this fact, that a *gath* cannot be defined except in terms of a *sam,* it follows that if by a *collection* be meant, as ordinarily is meant, a *gath;* while a *gath* is not distinguished from a *sam,* it becomes utterly impossible to define what is meant by a collection.

This would not be true if the two clauses of the definition of the *sam* were two distinct ideas which have to be put together; but it is not so. Secondness involves Firstness, although it can be discriminated from it; and consequently the idea of the existence of that which has an essence, which is simple Secondness, is a decidedly simpler notion than that of existence without essence, or a Secondness discriminated from Firstness. For it is only by a rectification applied to the former notion that the latter can be attained. No doubt the easiest way to conceive of the *sam* is to imagine that you have a common noun, without specifying what noun it is, and to think that that noun signifies some quality which is possessed by anything to which it applies, but is not possessed by anything to which it does not apply. Now you are to imagine a single thing which is composed of parts. Nothing is done to these parts to put them into their places in the whole: their mere existence locates them in the whole. Now think of this rule as describing the whole. If any individual object can properly have that common noun predicated of it, it is a part of the single object called a *sam;* if not, it is not. That gives you the idea of the *sam.* Now to get the idea of a *gath,* you are to consider that those individual objects might change their qualities without losing their individual identity; so that limiting ourselves to any instant any individual object which at that instant forms a part of the *sam* forever forms a part of an object of which no object not at that instant a part of the sam is a part, and this individual composite whole which has nothing to do with the qualities of its members is a *gath.*

For every *gath* there must be a corresponding *sam.* This is what we should ordinarily express by saying that whatever exists is possible. Or, as De Morgan put it, the individuals of whatsoever collection have some quality common to them all that is peculiar to them i.e. possessed by nothing else.[3] Kant, I dare say, would remark that this is a Regulative principle but that it cannot be proved to be a Constitutive principle. That is, it is proper to assume it, but you cannot prove it is so.[4] But I reply that every principle of logic is a Regulative Principle and nothing more. Logic has nothing to do with Existence. And I should add: Herr Professor Dr. Hofrath Kant, permit me to say that in saying this is not a Constitutive principle you speak of *qualities* as if they were existent individuals. A quality has no other being in itself than possibility and to say that a quality is possible is to say that it has all the being that in the nature of things a quality could have. If as you say there *may* be a quality common and peculiar to all the members of a gath, then there certainly is such a quality; and you yourself have in this very same breath described one such quality, in saying that they are all members of the gath in question. So for every gath there is a corresponding *sam.* But it is not true that for every *sam* there is a corresponding *gath.* Since there is the *sam* of the phenix, although it happens not to exist up to date. But there is no such *gath* since there is no phenix. Another point which I observe puzzles the Hon. Bertrand Russell in his 'Principles of Mathematics' is whether a collection which has but a single individual member is identical with that individual or not.[5] The proper answer is that if by a *collection* you mean a *sam,* the *sam* of the sun is not the sun, since it is an *ens rationis* having an essence, while the individual has no essence and is not an *ens rationis.* But if you mean the *gath,* the gath of the sun has no being at all except the existence of the sun which is all the being the individual existent sun has. Therefore, having precisely the same being they are identical and no distinction except a grammatical or linguistic one can be drawn between them. Mr. Russell's being puzzled by this is a good illustration of how impossible it is to treat of philosophy without making a special vocabulary such as all other sciences make. It is, however, far more needed in philosophy than in any other science, for the reason that the words of ordinary speech are needed by philosophy for its raw material.

What has been said of qualities is equally true of relations, which may be regarded as the qualities of sets of individuals. That is to say, if any form of relation is logically possible between the members of two given *gaths,* a relation of that form actually exists between them.

14

The Logic of Quantity

[Peirce 1893c] Kant and Mill, the principal objects of Frege's critical scrutiny in *The Foundations of Arithmetic,* were of great importance for Peirce as well, and in this selection he locates his epistemology for mathematics relative to theirs.[1] He takes Kant to task for holding that analytic truths can be ascertained by "a simple mental stare," and proceeds to rewrite Kant along the future-oriented lines of his own pragmatism: to find out what is *in*volved in a concept, we must see what we can *e*volve from it by way of experimentation on diagrams. As in the somewhat later selection 7, Peirce uses a modified associationist framework to unify the inner experimentation of the mathematician and the outer experimentation of the natural scientist. He acknowledges that "the difference between the inward and outward worlds is very, very great, with a remarkable absence of intermediate phenomena"; but ultimately the difference is "merely one of how much." This reconstruction of the analytic/synthetic distinction enables Peirce to put mathematics, in opposition to Kant, on the analytic side.[2] His way of doing this looks, at first, like a kind of logicism, but he holds in the end to his usual view of mathematics as "prelogical." So he is not saying that mathematics (in particular, arithmetic) rests on logic, but rather that (successful) mathematical reasoning *does* unfold what is "involved" in its hypotheses: this is just a quasi-Kantian way of saying that mathematics is "the science which draws necessary conclusions."

Yet mathematics is also an experimental science, and Peirce agrees with Mill that "*experience* is the only source of any kind of knowledge." At the same time he denies that mathematics is experiential in Mill's sense. He concedes that putative counterexamples to simple arithmetical propositions are conceivable; indeed, he holds that they "often happen" but are not genuine counterexamples because the "arithmetical propositions are not understood in an experiential sense." Their justification in inner experience renders them immune to the kind of refutations Mill envisions. But that justification does nonetheless involve experience broadly speaking, and so Mill is wrong to say that logically necessary propositions are, by virtue of their necessity alone, merely verbal. Peirce refuses to call them *a priori* because this suggests that their discovery is a matter of "[applying] plain rules to plain cases." He prefers to call them *innate,* "because that may be innate which is very abstruse, and which we can only find out with extreme difficulty." In the final analysis, then, Peirce holds that arithmetic is analytic yet experimental, and neither *a priori* nor *a posteriori,* but innate. It may be doubted whether he really has found a "third way" between Kant and Mill. No doubt Frege would have felt that Peirce took too much from them both. In any case this selection bears close comparison with Frege's more famous critique.

§231. Kant, in the Introduction to his Critic of the Pure Reason,[3] started an extremely important question about the logic of mathematics. He begins by drawing a famous distinction, as follows:

> In judgments wherein the relation of a subject to a predicate is thought . . . this relation may be of two kinds. Either the predicate, B, belongs to the subject, A, as something covertly contained in A as a concept; or B is external to A, though connected with it. In the former case, I term the judgment analytical; in the latter synthetical. Analytical judgments, then, are those in which the connection of the predicate with the subject is thought to consist in identity, while those in which this connection is thought without identity, are to be called synthetical judgments. The former may also be called explicative, the latter ampliative judgments, since those by their predicates add nothing to the concept of the subject, which is only divided by analysis into partial concepts that were already thought in it, though confusedly; while these add to the concept of the subject a predicate not thought in it at all, and not to be extracted from it by any analysis. For instance, if I say all Bodies are extended, this is an analytical judgment. For I need not go out of the conception I attach to the word *body*, to find extension joined to it; it is enough to analyze my meaning, i.e. merely to become aware of the various things I always think in it, to find that predicate among them. On the other hand, if I say, all bodies are heavy, that predicate is quite another matter from anything I think in the mere concept of a body in general.

Like much of Kant's thought this is acute and rests on a solid basis, too; and yet is seriously inaccurate. The first criticism to be made upon it is, that it confuses together a question of psychology with a question of logic, and that most disadvantageously; for on the question of psychology, there is hardly any room for anybody to maintain Kant right. Kant reasons as if, in our thoughts, we made logical definitions of things we reason about! How grotesquely this misrepresents the facts, is shown by this, that there are thousands of people who, believing in the atoms of Boscovich, do not hold bodies to occupy any space.[4] Yet it never occurred to them, or to anybody, that they did not believe in corporeal substance. It is only the scientific man, and the logician who makes definitions, or cares for them.

§232. At the same time, the unscientific, as well as the scientific, frequently have occasion to ask whether something is consistent with their own or somebody's meaning; and that sort of question they themselves widely separate from a question of how experience, past or possible, is qualified. The Aristotelian logic,—and, in fact, all men who ever have thought, have made that distinction. It is embodied in the conjugations of some Barbarous languages. What was peculiar to Kant,—it came from his thin study of syllogistic figure,—was his way of putting the distinction, when he says we necessarily think the explicatory proposition, although confusedly, whenever we think its subject. This is monstrous! The question whether a given thing is consistent with a hypothesis, is the question of whether they are logically compossible or not. I can easily throw all the axioms of number, which are neither numerous nor complicated, into the antecedent of a proposition,—or into its *subject*, if that be insisted upon,—so that the question of whether every number is the sum of three cubes, is simply a question of whether that is *involved* in the conception of the subject, and nothing more. But to say that because the answer is *involved* in the conception of the subject, it is confusedly *thought* in it, is a great error. To be *involved,* is a phrase to which nobody before Kant ever gave such a psychological meaning. Everything is involved which can be evolved. But how does this evolution of necessary consequences take place? We can answer for ourselves after having worked a while in the logic of relatives. It is not by a simple mental stare, or strain of the mental vision. It is by manipulating on paper, or in the fancy, formulae or other diagrams,—experimenting on them, *experiencing* the thing. Such experience alone *evolves* the reason hidden within us, and as utterly hidden as gold ten feet below ground; and this experience only differs from what usually carries that name in that it brings out the reason hidden within, and not the reason of Nature, as do the chemist's or physicist's experiments.

§233. There is an immense distinction between the Inward and the Outward truth. I know them alike by experimentation only. But the distinction lies in this, that I can glut myself with experiments in the one case, while I find it most troublesome to obtain any that are satisfactory in the other. Over the Inward, I have considerable control, over the Outward very little. It is a question of degree only. Phenomena that inward force puts together appear *similar;* phenomena that outward force puts together appear *contiguous.* We can try experiments establishing *similarity* so easily, that it seems as if we could see through and through that; while *contiguity* strikes us as a marvel. The young chemist precipitates prussian blue from two nearly colorless fluids a hundred times over without ceasing to marvel at it. Yet he finds no marvel in the fact that any one precipitate when compared in color with the sky seems similar every time. It is quite as much a mystery, in truth, and you can no more get at the heart of it, than you can get at the heart of an onion.

But nothing could be more extravagant than to jump to the conclusion that because the distinction between the Inward and the Outward is merely one of how much, therefore it is unimportant; for the distinction between the unimportant and the important is itself purely one of little and much. Now, the difference between the inward and the outward worlds is certainly very, very great, with a remarkable absence of intermediate phenomena.

§234. The first question, then, to ask concerning arithmetical and geometrical propositions is, whether they are logically necessary and merely relate to hypotheses, or whether they are logically contingent and relate to experiential fact.

Beginning with the propositions of arithmetic, we have seen already that arithmetical propositions may be syllogistic conclusions from ordinary particular propositions. From

$$A\overline{\overline{B}}$$

$$\text{and } \overline{A}\,\overline{\overline{B}}$$

Taken together, or

$$\text{Some}\,A \text{ is } B$$

$$\text{Some not-}A \text{ is } B$$

It follows that there are at least two *B*'s. This inference is strictly logical, depending on the principle of contradiction, that is, on the non-identity of *A* and not-*A*. By the same principle, from

$$\text{Some } A \text{ is } B,$$

$$\text{Some not-}A \text{ is } B,$$

$$\text{Any}\,B \text{ is } C,$$

$$\text{Some not-}B \text{ is } C,$$

taken together, it follows that there are at least three *C*'s.

Hamilton admits that the arithmetical proposition, "Some *B* is not some-*B*," is so urgently called for in logic, that a special propositional form must be made for it. So, if a distributive meaning be given to "every," Every *A* is every *A*, implies that there is but one *A*, at most. This is what this proposition must mean, if it is to be the precise contradictory of the other. If a proposition is infra-logical in form, its denial must be admitted to be so.[5]

It clearly belongs to logic to evolve the consequences of its own forms. Hence, the whole of the theory of numbers belongs to logic; or rather, it would do so, were it not, as pure mathematics, *prelogical,* that is, even more abstract than logic.

§235. These considerations are sufficient of themselves to refute Kant's doctrine that the propositions of arithmetic are "synthetical." As for the argument of J.S. Mill, or what is usually attributed to him, for what this elusive writer really meant, if he precisely meant anything, about any difficult point, it is utterly impossible to determine,—I mean the argument that, because we can conceive of a world in which when two things were put together, a third should spring up, therefore arithmetical propositions are experiential, this argument proves too much.[6] For, in the existing world, this often happens; and that fact that nobody dreams of its constituting any infringement of the truths of arithmetic shows that arithmetical propositions are not understood in any experiential sense.

But Mill is wrong in supposing that those who maintain that arithmetical propositions are logically necessary, are therein *ipso facto* saying that they are verbal in their nature. This is only the same old idea that Barbara in all its simplicity represents all there is to necessary reasoning, utterly overlooking the construction of a diagram, the mental experimentation, and the surprizing novelty of many deductive discoveries.

If Mill wishes me to admit that *experience* is the only source of any kind of knowledge, I grant it at once, provided only that by experience he means *personal history,* life. But if he wants me to admit that inner experience is nothing, and that nothing of moment is found out by diagrams, he asks what cannot be granted.

The very word *a priori* involves the mistaken notion that the operations of demonstrative reasoning are nothing but applications of plain rules to plain cases. The really unobjectionable word is *innate;* for that may be innate which is very abstruse, and which we can only find out with extreme difficulty. All those Cartesians who advocated innate ideas took this ground; and only Locke failed to see that learning something from experience, and having been fully aware of it since birth, did not exhaust all possibilities.

Kant declares that the question of his great work is "How are synthetical judgments *a priori* possible?"[7] By *a priori* he means universal; by synthetical, experiential (i.e. relating to experience, not necessarily derived wholly from experience.) The true question for him should have been, "how are universal propositions relating to experience to be justified?" But let me not be understood to speak with anything less than profound and almost unparalleled admiration for that wonderful achievement, that indispensible stepping stone of philosophy.

15

Recreations in Reasoning

[Peirce 1897(?)b] Along with selection 10, this mathematico-philosophical treatment of the natural numbers is one of the strongest pieces of evidence in support of the claim that Peirce held to a kind of mathematical structuralism.[1] It is also, as discussed in the introduction (xxxix), an important source of information about Peirce's views on the metaphysics of mathematical structure.

The selection culminates in a Dedekindian axiomatization of the natural numbers, and the derivation of some fundamental properties—including the Fermatian Principle, or mathematical induction—from the basic axioms.[2] But Peirce's first approach to numbers here is semiotic. A number is, in the first place, a "meaningless vocable" used in counting collections. Numerals are familiar examples of numbers in this primary sense, but so are some nonsense syllables from children's games. Such numbers, recited in a standard order, are used to run through a collection; the one that exhausts the collection then functions as an adjective expressive of that objective attribute of the collection which Peirce calls its "multitude, or collectional quantity."

When Peirce says that numbers are "meaningless" he is in effect classifying them semiotically as indices; in Mill's terminology, they are signs with denotation only and no connotation. He effects the classification here against the background of the associationist account of the inner and outer worlds he used also in selection 7. Here, as there, the associationist language masks his categories and his semiotics (though Thirdness comes through much more clearly here than there). But in this context it also provides a vocabulary in which he can couch his account of abstract number.

The function of indices is to point things out, and if numbers are indices it is fair to ask what it is they point to. Peirce neither poses nor answers that question directly, but a natural answer is that each number points to a position in the order of counting. We can think of the theory of abstract number which Peirce develops in the second half of the selection as the theory of these positions, viewed in abstraction from the numbers (vocables) that indicate them. Peirce does not make systematic use of hypostatic abstraction in explaining how the concept of number, or the conception of numbers as objects, arises from the practice of counting. But abstraction is arguably implicit in the transition from meaningless vocables to adjectives, and in

Peirce's classification of "multitude . . . [as] an *attribute* . . . of collections" [emphasis added]. This is one of the many avenues for philosophical exploration opened up by this richly suggestive text.

When a number is mentioned, I grant that the idea of a succession, or transitive relation, is conveyed to the mind; and in so far the number is not a meaningless vocable. But then, so is this same idea suggested by the children's gibberish

"Eeny, meeny, mony, mi."

Yet all the world calls these meaningless words, and rightly so. Some persons would even deny to them the title of "words," thinking, perhaps, that every word proper means something. That, however, is going too far. For not only "this" and "that," but all proper names, including such words as "yard" and "metre" (which are strictly the names of individual prototype standards), and even "I" and "you," together with various other words, are equally devoid of what Stuart Mill calls "connotation." Mr. Charles Leland informs us that "eeny, meeny," etc. are gipsy numerals.[3] They are certainly employed in counting nearly as the cardinal numbers are employed. The only essential difference is, that the children count on to the end of the series of vocables round and round the ring of objects counted; while the process of counting a collection is brought to an end exclusively by the exhaustion of the collection, to which thereafter the last numeral word used is applied as an adjective. This adjective thus expresses nothing more than the relation of the collection to the series of vocables.

Still, there is a real fact of great importance about the collection itself which is at once deducible from that relation, namely, that the collection cannot be in a one-to-one correspondence with any collection to which is applicable an adjective derived from a subsequent vocable but only to a part of it; nor can any collection to which is applicable an adjective derived from a preceding collection[4] be in a one-to-one correspondence with this collection, but only with a part of it; while on the other hand this collection is in one-to-one correspondence with every collection to which the same numeral adjective is applicable. This, however, is not essentially implied as a part of the significance of the adjective. On the contrary, it is only shown by means of a theorem, called "The Fundamental Theorem of Arithmetic," that this is an attribute of the collections themselves and not an accident of the particular way in which they have been counted. Nevertheless, this is a complete justification for the statement that quantity,—in this case, multitude, or collectional quantity,—is an attribute of the collections themselves. I do not think of denying this; nor do I mean that any kind of quantity is merely subjective. I am simply not using the word quantity in that acception. I am not speaking of physical, but of mathematical, quantity.

Were I to undertake to establish the correctness of my statement that the cardinal numerals are without meaning, I should unavoidably be led into a disquisition upon the nature of language quite astray from my present purpose. I will only hint at what my defence of the statement would be by saying that, according to my view, there are three categories of being, ideas of feelings, acts of reaction, and habits. Habits are either habits about ideas of feelings or habits about acts of reaction. The ensemble of all habits about ideas of feeling constitutes one great habit, which is a World; and the ensemble of all habits about acts of reaction constitutes a second great habit, which is another World. The former is the Inner World, the world of Plato's forms. The other is the Outer World or universe of existence. The mind of man is adapted to the reality of being. Accordingly, there are two modes of association of ideas, inner association, based on the habits of the inner world, and outer association, based on the habits of the universe. The former is commonly called association by resemblance; but in my opinion, it is not the resemblance which causes the association, but the association which constitutes the resemblance. An idea of a feeling is such as it is within itself, without any elements or relations. One shade of red does not in itself resemble another shade of red. Indeed, when we speak of a shade of red, it is already not the idea of the feeling of which we are speaking but of a cluster of such ideas. It is their clustering together in the Inner World that constitutes what we apprehend and name as their resemblance. Our minds being considerably adapted to the inner world the ideas of feelings attract one another in our minds, and in the course of our experience of the inner world develope general concepts. What we call sensible qualities are such clusters. Associations of our thoughts based on the habits of acts of reaction are called associations by contiguity, an expression with which I will not quarrel, since nothing can be contiguous but acts of reaction. For to be contiguous means to be near in space at one time; and nothing can crowd a place for itself but an act of reaction. The mind, by its instinctive adaptation to the Outer World, represents things as being in space, which is its intuitive representation of the clustering of reactions. What we call a thing is a cluster or habit of reactions, or, to use a more familiar phrase, is a centre of forces. In consequence, of this double mode of association of ideas, when man comes to form a language, he makes words of two classes, words which denominate things, which things he identifies by the clustering of their reactions, and such words are proper names, and words which signify, or *mean,* qualities, which are composite photographs of ideas of feelings, and such words are verbs or portions of verbs, such as are adjectives, common nouns, etc.

Thus, the cardinal numerals in being called meaningless are only assigned to one of the two main divisions of words. But within this great class the cardinal numerals possess the unique distinction of being mere instruments of experimentation. "This" and "that" are words designed to stimulate the person addressed to perform an act of observation; and many other words have that character; but these words afford no particular help in making the observation. At any rate, any such use is quite secondary. But the sole uses of the cardinal numbers, are, first, to count with them, and second to state the results of such counts.

Of course, it is impossible to count anything but clusters of acts, i.e. events and things (including persons); for nothing but reaction-acts are individual and discrete. To attempt, for example, to count all possible shades of red would be futile. True, we count the notes of the gamut; but they are not all possible pitches, but are merely those that are customarily used in music, that is, are but habits of action. But the system of numerals having been developed during the formative period of language, are taken up by the mathematician, who generalizing upon them creates for himself an ideal system after the following precepts.

PRECEPTS FOR THE CONSTRUCTION OF THE SYSTEM OF ABSTRACT
NUMBERS.

1st, There is a relation, G, such that to every *number,* i.e. to every object of the system a different number is G and is G to that number alone; and we may say that a number to which another is G is "G'd by" that other;

2nd, There is a number, called zero, 0, which is G to no cardinal number;

3rd, The system contains no object that it is not necessitated to contain by the first two precepts. That is to say, a given description of number only exists provided the first two precepts require the existence of a number which may be of that description.

This system is a cluster of ideas of individual things; but it is not a cluster of real things. It thus belongs to the world of ideas, or Inner World. Nor does the mathematician, though he "creates the idea *for himself*," create it absolutely. Whatever it may contain of impertinent is soilure from. The idea in its purity is an eternal being of the Inner World.

This idea of discrete quantity having an absolute minimum subsequently suggests the ideas of other systems, all of which are characterized by the prominence of transitive relations. These mathematical ideas being then applied in physics to such phenomena as present analogous relations form the bases of systems of measurement. Throughout them all, succession is the prominent relation; and all measurement is effected by two operations. The first is the experiment of superposition the result of which is that we say of two objects, A and B, A is (or is not) in the transitive relation, τ, to B, and B is (or is not) in the relation τ to A; while the second operation is the experiment of counting. The question "How much is A?" only calls for the statement, A has the understood transitive relation to such things, and such things have this relation to A.

APPLICATION TO THE THEORY OF ARITHMETIC.

According to the theory partially stated above, pure arithmetic has nothing to do with the so-called Fundamental Theorem of Arithmetic. For that theorem is that a finite collection counts up to the same number in whatever order the individuals of it are counted. But pure arithmetic considers only the numbers themselves and not the application of them to counting.

In order to illustrate the theory, I will show how the leading elementary propositions of pure arithmetic are deduced, and how it is subsequently applied to counting collections.

Corollary 1. No number is G of more than one number. For every number necessitated by the first precept is G to a single number, and the only number necessitated by the second precept, by itself, is G to no number. Hence, by the third precept, there is no number that is G to two numbers.

Corollary 2. No number is G'd by two numbers. For were there a number to which two numbers were G, one of the latter could be destroyed without any violation of the first two precepts, since the destruction would leave no number without a G which before had one, nor would it destroy 0, since that is not G. Hence, by the third precept, there is no number which is G to a number to which another number is G.

Corollary 3. No number is G to itself. For every number necessitated by the first precept is G to a different number, and to that alone; and the only number necessitated by the second precept, by itself, is G to no number.

Corollary 4. Every number except zero is G of a number. For every number necessitated by the first precept is so, and the only number directly necessitated by the second is zero.

Corollary 5. There is no class of numbers everyone of which is G of a number of that class. For were there such a class, it could be entirely destroyed without conflict with precepts 1 and 2. For such destruction could only conflict with the first precept if it destroyed the number that was G to a number without destroying the latter. But no number of such a class could be G of any number out of the class by the first corollary. Nor could zero, the only number required to exist by the second precept alone belong to this class, since zero is G to no number. Therefore, there would be no conflict with the first two precepts, and by the third precept such a class does not exist.

The truly fundamental theorem of pure arithmetic is not the proposition usually so called, but is the Fermatian principle, which is as follows:

> *Theorem I.* The Fermatian Principle. *Whatever character belongs to zero and also belongs to every number that is G of a number to which it belongs, belongs to all numbers.*

Proof. For were there any numbers which did not possess that character, their destruction could not conflict with the first precept, since by hypothesis no number without that character is G to a number with it. Nor would their destruction conflict with the second precept directly, since by hypothesis zero is not one of the numbers which would be destroyed. Hence, by the third precept, there are no numbers without the character.

16

Topical Geometry

[Peirce 1904(?)b] Peirce took a serious interest in what even in his own day was known as topology: he worked hard on the problem of map coloring, and also on what he held to be an improved formulation of Listing's Census Theorem.[1] The title of this text, intended for publication in *Popular Science Monthly,* is one of Peirce's preferred designations for topology; he introduces another—*geometrical topics*—in his opening paragraph. These names fit more smoothly than 'topology' does into a systematic nomenclature for the major branches of geometry; Peirce opens his article by explaining the system behind the nomenclature. Topics studies those properties of geometrical objects which are invariant under continuous motion and deformation. This definition of topics, in terms of invariant properties under certain transformations, is very close to that given by Felix Klein (1849–1925) in his Erlangen Programme.[2] However, Peirce does not follow Klein in basing his definitions of the other branches of geometry on invariance of this sort.

Since "the only general concept of space we . . . can have, is as it is a law imposed upon certain changes of objects, namely, their motions," topics studies space itself; and since, unlike the other branches of geometry, it involves no experiential concepts, it is the mathematical theory of *pure* space. *Graphics* (projective geometry) makes use of the physically defined concept of straightness, and *Metrics*—the familiar geometry of Euclid's *Elements*—must add the further concept of a rigid body in order to make its quantitative comparisons of lengths and angles. This account of the branches of geometry draws on work by Klein and Arthur Cayley (1821–1895). Peirce alludes to Klein's contributions in selection 23 (181), but he only names Cayley here. In summarizing these results, it will be useful to proceed the other way around. Klein (1871) shows how to define a metric function, measuring segments and angles, on suitably chosen subsets of the complex projective plane, relative to suitably chosen conics in that plane. Depending upon the choice of conic, one obtains different metrical geometries. If the conic is real, one obtains what Klein calls *hyperbolic* geometry, in which the sum of the angles of a triangle is less than two right angles; if the conic is purely imaginary, one obtains *elliptic* geometry, in which that sum is greater than two right angles; and finally, if one chooses a particular kind of degenerate conic, what results is *parabolic* (that is, Euclidean) geometry, in which the sum is equal to two right angles. Cayley (1859) had obtained similar but more limited results in the affine plane, but he did not consider the case in which the conic (which Cayley called the Absolute) is real.[3] He drew two dramatic conclusions from these discoveries (592), both of which profoundly influenced Peirce's philosophy of geometry. The first is "that the metrical properties of a figure are not the properties of the figure *per se* apart from everything else, but its properties when considered in connexion with another figure, viz. the conic termed the Absolute." The second is that "metrical geometry is . . . a part of descriptive [i.e., projective] geometry, and descriptive geometry is *all* geometry." Peirce's translation of Cayley's second conclusion into his own vocabulary is that "metrics is, at bottom, only a particular problem of graphics"; he goes on to argue that, likewise, "the graphics of *real* Space (i.e. space without 'imaginary' parts) is but a particular problem of topics."

Motion takes place in time, so time is presupposed by topical geometry. Topical time, like topical space, is more general than experienced time, the most notable difference being that topical time has no preferred direction. However it is, like time on our common-sense conception of it, a true continuum. Peirce sharply distinguishes this true continuity, the continuity of "topicists" like himself, from the "pseudo-continuity" of "the analysts." A true continuum is not, as writers like Cantor and Dedekind would have it, an aggregate of indivisible elements which is order-isomorphic to the real numbers. Indeed, a true continuum contains no indivisible elements at all. Peirce argues this point in a long footnote making amends for his treatment of continuity in selection 19. Though he faults both himself and Kant for confusing continuity and infinite divisibility, his argument against indivisibles rests on Kant's account of the continuity of time, "as consisting in every part of it being itself a lapse of time." An instant would be an indivisible part of time, that is, a part of time which was *not* a lapse. So if time is continuous in Kant's sense, it contains no instants.

In the closing paragraphs of this selection (which do not conclude the manuscript from which it is taken) Peirce deals summarily with two major themes in his philosophy of continuity: its experiential presence, and the "supermultitudinousness" of a true continuum. On the first point, he considers a "pragmatistic" objection to the very concept of a true continuum: since the concept can make no difference to conduct, it is not a *definite* concept. His reply draws on his late conception of logic as a normative science which evaluates self-controlled processes of reasoning. All of our "introspective perceptions" put us in a position to judge that time is continous; this latter judgment is not self-controlled and hence not open to logical criticism. This hardly answers the pragmatistic worry about conduct, but the idea that we have something like a transcendental grasp of continuity is one Peirce kept coming back to in his writings on the subject.

Peirce puts forth the supermultitudinousness of truly continuous time as a feature which distinguishes it from analytical "pseudo-continuity." We can always find room, in a true continuum, for an arbitrarily large collection of points. A pseudo-continuum is composed of points; so its points constitute a definite collection with a well-defined multitude, and leave room for no additional points. But as Peirce proves in a passage omitted here, for any well-defined multitude there is a greater one. (For a proof, see selection 22.) So a collection of points whose multitude exceeds that of the pseudo-continuum *P* is both possible and incapable of fitting into *P*. Hence *P* is not supermultitudinous, as a true continuum is.

The Popular Science Monthly has, of late years, done such incalculable good both to those who read it and to those who do not, by recording the advances of scientific ideas, that I, for one, feel it to be a definite duty, when summoned to contribute, to do what one can in response to the call.

The Three Main Departments of Geometry.[4] Topology, or, as I prefer to call it, topical geometry, or, still better, *geometrical topics,* is a subject concerning which everybody ought to know, though few do, the little that has ever been made out. It is the most fundamental and, at the same time, the simplest of the three great divisions of geometry,

> *topics,*
> *graphics,*
> and *metrics.*

Metrics, what. Metrics *embraces* the science of all spatial quantities. The very first spatial quantity that presents itself is the relative length of a line; the second is the relative magnitude of an angle. In terms of these all other quantities are definable. But if we suppose Space to be everywhere perfectly continuous, and, as such, entirely smooth and homogeneous it can possess, *in itself,* in its pure vacuity, no characters by which the relative lengths of two lines or the relative magnitudes of angles could be determined. For the determination of those magnitudes, we are compelled to refer to the displacements of an imaginary rigid body. If, as a partial definition of equality, we agree to call the places that can at different times be occupied by this body *equal,* and if we then extend the meaning of geometrical quantity by suitable "axioms" of quantity, so as to make it applicable to sums, differences, limits of endless series, and perverse figures, there is nothing more to prevent our quantitative comparison of those homogeneous geometrical characters whose variations take place in linear series. Since the whole doctrine rests upon nothing except the formal "axioms" and the properties of the rigid body, metrics is best defined as the science of the geometry of the rigid body. It certainly is not a doctrine about pure Space itself; and it is best to exclude from it the doctrine [of] the kind of quantity,—especially, the cross-ratio,[5]—that does not depend upon the properties of the rigid body.

Graphics, what. Graphics, or projective geometry, is simply an extension of the doctrine of linear perspective, resulting from supposing the lines of sight to be unlimited rays that do not stop at the eye, but pass through it; so that the picture shows what is behind the spectator, as well as what is in front of him. Straight lines have no pure spatial properties which distinguish them from other families of lines in Space. They might be defined as the paths of freely moving particles not subjected to any forces. In any case, they are lines subject to a general condition incapable of definition in terms of Space itself. Therefore, graphics is no more a branch of pure geometry than is metrics.

Topics, what. With topics, however, the case is different. In this field of thought we still suppose objects to move about in Space. But we suppose that, at will, any of these objects can be made to expand, to contract, to bend, to twist, and in sort to move free from any law, excepting only that it is nowhere to be broken or welded;—or, to state the condition precisely, that no two parts or limits of it shall [be able] at one instant to occupy one and the same place and at another instant separate places. In the usual phrase of the mathematicians, we suppose the connection of parts to be undisturbed. Now this connection of parts, which is the sole law of topical movables, is a property of the very Space itself. Besides, the topicist holds himself at liberty to suppose even this law of connection to be broken, provided the violation be explicitly supposed to take place on a definite occasion and by a defined motion. Now mathematics, dealing as it does with purely hypothetical objects, can know no experiences or peculiar feelings. Space, therefore, for this science, can only be a general concept; and the only general concept of Space we have, or can have, is as it is a law imposed upon certain changes of objects, namely, their motions. Topics is thus the only mathematics of pure Space that is possible.

Mutual relations of the three geometries. It is a celebrated truth, familiar to mathematicians that metrics is, at bottom, only a particular problem of graphics; that whatever is true of spatial magnitudes consists in a graphical truth,—a truth of intersections and tangencies,—about one peculiar individual place, or *locus,* called the Absolute. Just in that form of statement, this only holds of the Space of analytical geometry, with its "imaginary" points and "imaginary" planes; but it unquestionably indicates a relation between Metrics and Graphics which subsists, in a modified way, even in "real" Space. This truth is a recondite one only brought to light by Cayley in 1858. It is, on the other hand, almost self evident that the graphics of *real* Space (i.e. space without "imaginary" parts) is but a particular problem of topics. For suppose all space to be filled with a fluid, not elastic but expansible and compressible, and incapable of any discontinuous motion. All the truths of graphics are resolvable into truths about the intersections of rays, or limitless straight lines. Every ray in Space will be occupied at any instant by a line of particles,—call it a *filament,*—of that fluid. Imagine any number of these filaments to be colored at any instant so as to be identifiable. Now let that fluid move in any way, no matter how intricate, and afterward to come to rest again. Those filaments will no longer be straight; for the simple expansion of one part of the fluid with a contraction of another part would suffice to destroy their straightness. Heaven only knows into what a snarl they might not be. Yet they would all intersect in the very same particles in which they intersected at the beginning of the motion. This shows clearly that rays cannot be distinguished from various other families of lines by any graphical properties. The whole family of rays is simply a particular object in Space precisely like innumerable others, just as the Absolute is a particular object in space precisely like innumerable others; and the doctrine of the intersections of rays is included under the general doctrine of the possible displacements of such a fluid, which latter doctrine is nothing but a particular problem of the general science of topics.

The Space of topics. For the Space of topics, be it understood, is not the Space of experience; for whatever is to be an object of pure mathematical reasoning must be a purely hypothetical object whose properties are virtually known to perfection, as those of no object of experience ever can be. There are many branches of topics. There is a topics of a Space of one dimension, a topics of a Space of any whole number of dimensions. Nor is that all. A Space of any number of dimensions greater than one may have different shapes, although it be perfectly continuous in every part. To this matter we shall have occasion to return. The present paper will be confined mainly to a Space of three dimensions, perfectly continuous, and of almost the simplest possible topical shape, a shape which we may suppose the Space of experience would have if its shape could be determinate. What shape may best be chosen we shall consider below. As long as it is contrary to the nature of experience that all parts of experiential Space should be known, it is pretty obvious that experiential Space cannot have a perfectly determinate shape. But it conveys a more accurate notion to say that we cannot know exactly what its shape may be.

Topics presupposes the doctrine of Time. The doctrine of topics presupposes the doctrine of Time, because it considers motions. I must therefore prefix to what I have to say about topics an explanation of the topical properties of Time.

Our Time has a psychical and a physical ingredient; the Time of topics has the latter only. Now the Time of our knowledge, considered as a system of relations between instants, has two different ingredients. Almost anybody would admit readily enough that the striking difference of aspect between the past and the future is quite illusory, the different dates themselves being all alike in themselves. It will not be so readily granted that the relation of a previous to a subsequent time, although opposite to the relation of the subsequent to the previous, nevertheless is exactly like it, just as the two opposite ways of passing along a spatial line differ in no respect from one another, although (or except that) they are the reverse each of the other. I personally believe that the two directions of Time are as alike as the two directions along a line. For the law of the conservation of energy is that the *vis viva,* and consequently also the forces, of particles depends upon nothing mutable except the relative positions of the particles. Now the differential of the time enters into the analytical expression of the *vis viva,*

$$\frac{1}{2}m\left(\frac{ds}{dt}\right)^2,$$

as well as into that of force,

$$m\frac{d^2s}{(dt)^2},$$

only as squared. Whence, the square of a negative quantity being equal to that of the corresponding positive quantity, the two directions of time are indifferent as far as the action of the law of the conservation of energy goes. This seems to me to indicate that the difference of the two directions through time consists in a peculiar property of *psychical* events, and not to purely *physical* events, and *a fortiori* not to pure Time itself. At any rate, be that as it may, for all purposes of topics, there is no more difference between the two directions along Time,—the purely hypothetical Time of topics,—than between the two directions along a physical line.

True continuity *vs.* the pseudo-continuity of the calculus. The hypothetically defined Time of topics, like the undefined Time of common sense, is a true continuum of a single dimension. Now true continuity is one of those things concerning the nature of which knowledge has been bestowed upon babes and sucklings and has been almost denied to those who reflect and analyze. So long as we trust to common sense, the properties of a true continuum are a matter of course; but as soon as we undertake precisely to define our meaning in general terms, we flounder from quagmire into quicksand. The result of all the industry of cogitation on this subject of many mathematicians has been, so far, that there remains a serious dispute between those who approach the subject from the side of the calculus and those who, like me, approach the subject from the side of topics. The topicists say that what the analysts call continuity is not continuity at all: we call it pseudo-continuity. But it must not be supposed that it is a dispute about words merely. For the analysts say no more perfect continuity than theirs is possible; and that the descriptions of continuity given by topicists are nonsensical. I must confess that the attempt I made in the Monist, Vol. II, pp. 542 *et seqq.*[6] to define true continuity was a failure. What I there defined as continuity was nothing but the pseudo-continuity of the analysts.[*] But I hope I have accomplished something in my twelve years of subsequent reflection. I am now prepared to analyze true continuity and to show that the conception involves no contradiction and is different from that of the pseudo-continuity of the analysts; and I fully admit that, after my mistake of 1892, it is incumbent upon me to publish a full discussion of the point. Unfortunately, I lack the two kinds of quantity we are just now considering. I neither have the space for the full discussion here, nor do the necessities of bread and butter allow me the time for that any more than for communicating other results. I can here only give a partial explanation of true continuity.

---

[*]. That passage contains, besides, a second error, consisting in saying that Kant defines continuity as infinite divisibility. I was not, however, guilty of originating this misinterpretation. It is Kant himself who, while always defining the continuity of Time as consisting in every part of it being itself a lapse of time, constantly confounds this with an infinite divisibility like that of the series of rational fractions. But if there be any instants in Time, Time consists of those instants in their relation to one another; and therefore, under that condition, instants are parts of time. Now instants are indivisible; and consequently, to say that every part of Time is a lapse of time implies that there are no instants at all in Time, or at least not in uninterrupted Time.

Not a question of fact. Let me repeat that in topics we are in the domain of pure mathematics, whose business it is to study whatever definite hypotheses it may find interesting, without assuming the least responsibility for their accordance with fact. If God has not put the idea of true continuity into His World of Nature, it is, at any rate, to be found in the World of Ideas, which is equally His.

The pragmatistic objection. Yet, merely for the sake of repelling the objection that, according to my own doctrine of pragmatism, the conception of true continuity cannot be a definite conception, inasmuch as it has no conceivable bearing upon conduct, I will just mention, as a sufficient reply to the *argumentum ad hominem,* that I hold that direct judgments concerning present perceptions,—as when I say to myself, 'This seems red,'—are not open to logical criticism, because it is idle to blame what one cannot control, and that such a judgment concerning any introspective perception is, 'The connection of past and future seems truly continuous.' Thus, my own opinion is that true continuity is confusedly apprehended in the continuity of common sense. But this is aside from the question in hand, which is, whether it is possible to give any other definite meaning to continuity than that of the differential calculus and doctrine of limits.

The multitude of instants in pseudo-continuous Time. Can I mention any definite point of difference between the two continuities? Yes; among others this, that in a truly continuous Time there is room between any two instants for any multitude of instants whatsoever. The analysts assume that, besides instants at all measurable distances from any instant, adopted as an epoch from which to reckon time,—and meaning, of course, by "measurable," measurable by sufficiently increasing human powers of subdivision and of enumeration, without altering the nature of those powers,—there are also instants at "irrational" distances, that is, distances incapable of exact measurement and therefore incapable of precise designation in terms of measurement, but only describable as the results of series of computations, each computation being practicable, but the series not so because it is endless. Concerning such a description they make two assumptions; first, that there is an instant to which it is applicable; second, that the description distinguishes the one instant to which it applies from all others. Since these assumptions involve no contradiction, analysts have a perfect right, as pure mathematicians, to make them. I will now show that this constitutes a difference between analytical pseudo-continuity and true topical continuity, since, according to the second assumption, there is not room between any two instants of analytical Time for any multitude of instants whatsoever. For if

Time actually has instants all along its course, as the analysts say, then it consists of those instants in their relation to one another, and has no room for any other instants; and so the analysts expressly assert (G. Cantor in Vol XVII of the *Annalen*). . . .

Since . . . every multitude is exceeded by some other, it follows that the analyst's Time which has no room for other instants than those it actually contains, has no room for a greater multitude of instants than that, while there are greater multitudes. In this it differs distinctly from truly continuous Time, in which there is room for any multitude of instants whatsoever between any two instants. Since truly continuous Time has room for any multitude of instants whatsoever, in whatever sense those instants are contained in it they must be *welded together,* so as to lose their distinctness. For otherwise they would be a collection whose multitude would be exceeded by another possible multitude and thus there would not be room in Time for any multitude of instants whatsoever. The instants cannot all be present in time as instants.[*]

---

[*]. This was remarked by Herz, Kant's correspondent, in his *Betrachtungen* upon Kant's Inaugural Dissertation of 1770.

17

A Geometrico-Logical Discussion

[Peirce 1906(?)] The ontological status of points is a central, and trouble-some, issue in Peirce's philosophy of continuity. In this late text he attempts to address that issue in the setting of his general account of mathematical objects as *entia rationis.* Peirce introduces this term of art with a near-trans-lation ("creations of thought") that highlights its mentalistic overtones, but these give way, in the main discussion of points, to modal language that is more compatible with the less idealistic analyses of, e.g., selections 13 and 15, and also with his mature accounts of continuity.

Both points and collections, though created by acts of thought, have objective properties. A collection has a determinate number of members, about which the one whose thought creates the collection can be mistaken. In his discussion of points Peirce does not name a mind-independent property of a mental creation; indeed, the ontological thrust of his example would seem to be that points, lines, and surfaces are actually present, as in the boundaries between air, ice, water, and wood in a water bucket with a broken sheet of ice stuck to its side. We might try to bring such points under the heading of *entia rationis* by saying that they mark discontinuities that we might notice; their objectivity would consist in the mind-independence of the discontinuities. But if the discontinuities, along with the points that mark them, are there to be noticed, then they would seem to be independent objects of thought rather than creations of it.

Peirce goes on to consider points with a better claim to be *mere* cre-ations of thought, or rather, to be realities that are parasitic upon certain kinds of possibilities. He considers the two points where ice, air and water meet (these are the extremities of the line segment formed by the top surface of the ice and the side of the bucket). Tilt the bucket far enough and the water will completely submerge the surface of the ice where it sticks to the bucket; as Peirce picturesquely puts it, the two points we began with "run together" and then disappear. In a striking allusion to Meinong, Peirce says that "they both pass into the *Jenseits,*—the glorious realm of the Impossi-ble," but surely his meaning is that they become mere possibilities, mere *entia rationis,* by passing into the realm of the possible but unactualized; for if we tilt the bucket back, the points will reappear. Towards the end of the selection the connection with continuity becomes clear: if points are present in a continuum without being distinguished, even by an act of thought, the continuum will not be supermultitudinous. Peirce does not explain here why we should not resolve this tension by rejecting supermultitudinousness rather than by rejecting a fully realistic view of points.

The selection opens with a brilliant aside on Weierstrass's saw-tooth function which we can only wish Peirce had fleshed out more fully. The function is continuous everywhere and differentiable nowhere; Peirce mentions it in the first place as an example of mathematical surprise. But it is also more deeply problematic. Nondifferentiability everywhere implies that the graph of the function cannot be the path of a moving particle, because there is no definite direction, at any point on the curve, for the particle to move in. Peirce locates the problem in the use of completed infinite sums to define the function. Here, as in his ontology of points, Peirce flirts with constructivism. At the same time, as we will soon see, Peirce's theory of continuity (except in its very latest phase) comes to rest on the highly nonconstructive notion of an abnumeral (that is, uncountable) multitude. Along with (perhaps inseparably from) the ontology of points, this constructivist impulse sets up an enduring tension in Peirce's thinking about continuity.

Given 4 Rays, How many rays cut each all those four, under different circumstances? . . .

I know of no question which through equally facile steps leads to a better *aperçu* of the nature of Projective Geometry and, though less definitely, of the nature of mathematics in general, than does the question I have set down above. I therefore propose to give a discussion of it. In order to make the discussion complete and go to the bottom of the question, I should have to begin by expounding the logic of relatives, then going on successively to the doctrine of multitude, and the doctrine of time, and finally coming to Geometry. I do not, however, propose to go back to the cosmogony, or creation of the world (which "has puzzled the philosophers of all ages"). I shall just slightly allude to some things that may not be familiar to all and shall then go on to the strictly geometrical principles of the kind I propose to make use of without inquiring very closely whether each one will be wanted or not. I will begin with a few

1. Ordinary logic supposes us to be dealing with absolutely exact ideas. Yet absolute exactitude of thought is quite impossible, and out of the nature of things. Perhaps as exact an idea as we have is that of an individual person; and yet after all the observations that have been made upon multiple personality, and especially of the state of persons who are in process of cure of a state of changed personality, it must be confessed that the conception is only moderately definite. The idea of a species of animal, say a dog, is obviously quite nebulous upon its borders; and it is easy to imagine an animal that might happen to get called a dog or might just as likely get called a wolf, according to the accidental formation of one habit of thought or the other. Mathematical ideas are among the most, if not the very most, exact of all general ideas. Yet the surprises we have had in the last thirty years or so make us feel that they are far from being as unmistakable as we have imagined them to be. For example, it used to be accepted as a manifest truth that every line must have a definite direction at every point. Subsequently, Weierstrass and the general body of mathematicians denied that, because, they said, a line may be wavy in such a way that on every wave there are smaller waves, and *therefore* on these waves still smaller ones without end.[1]

It is undeniable that a line, once formed, can become crinkly; and we can in a general way admit that every crinkle can attain crinkles on it. There is, thus, no contradiction in thinking that this crinkling goes on endlessly; but whether it means anything to say that this endless crinkling is *completed* is not quite clear to me. A line is the path of a moving point. Now whether it means anything to say that a point starts to move without starting to move in any particular direction, is to me *more* than doubtful. I declare that there is no such idea at all. A *possibility* may be endless; but I cannot admit that an endless series ever actually gets ended. It may happen that actual fact is *represented* by an endless series. A ninth part of a dollar is possible. But I do not admit that it can be made by adding to a *dime* a *cent,* and then a *mill,* and then a tenth of a mill and so on endlessly. I think that when a point starts to move, it must start in some definite direction. For when something is *done,* it must be a definite thing that is done.

2. We know only too well that all things are not just as we would like to have them; and that the only way of improving the situation is to *do* something about it: mere dreaming will not answer the purpose. This is what is meant by saying that things are *real.* If the world were a dream, we could just dream otherwise. But the real is that which is as it is whatever you or I may think about it. Yet there are things that are certainly mere creations of thought, "*entia rationis,*" which, nevertheless, once created are quite real. For example, the Pleiades are a cluster of stars. It is the *fact* that makes them so. But if I choose to say that the city of New York, the precession of the equinoxes (or the perihelion of the orbit of Mars) and the tail of the wooden horse that was taken into Troy are a *trio,* it thereupon becomes true. They are neither two nor five nor any other number. They are three. But the only thing that makes them so is the fact that I took it into my head to think of them as a collection or plural. But there is a very subtle distinction of thought to be drawn here. Namely, it was not my thinking, as mere thinking, that made them a trio, but it was because my thinking was a real performance that actually *did* bring images of those things together. To show that this was so, suppose that instead of picking out three things I had picked out what I supposed to be thirty. Now suppose I had been mistaken and there were only twenty-nine. Then my thinking them to be thirty would not make them so. Another example of the same distinction, but much more convincing, and at the same time with more relation to questions that might puzzle one is the mathematical point. A point may be just as objective and empirical as anything in the world. You wake up in the morning and find the wooden pail of water in your room has a thin lamina of ice over it. You break it, but still a piece of it sticks to the side of the pail. In this condition quite as perceptible, or more so, than the four substances, wood, water, ice, and air, are the six surfaces, of which one is wood-water, as much one as the other; another is wood-ice; a third, water-ice; the fourth, fifth, and sixth are wood-air, water-air, ice-air. These six surfaces come together three by three in four lines; of which one is wood-water-ice, another is wood-water-air, another is wood-ice-air, and the fourth water-ice-air. These four lines all meet at both ends in two points which are both wood-water-ice-air. These points are just as real as anything can be. It is true that if you tilt your pail toward the side where the ice adheres to it, the two points will run together. For a single instant they will be one, and then they both pass into the *Jenseits,*—the glorious realm of the Impossible.[2] During this process one of the four lines, that of wood-ice-air

will shrink to nothing while the others, at the instant of the coalescence of the points, will become self-returning. It is the nature of real points to be paired. If a line is continuous and not interrupted by any points being marked upon it, no points exist upon it, except its two extremities. The proof of that is that it is the nature of existence to be definite and distinct. There is *room* upon every line, however short for any multitude of points all perfectly distinct. This "room" means possibility. But if these points for which there is room existed there there would be some definite multitude of them and no more, while in fact, no matter what multitude of distinct points be placed upon the line there will always be room for more. But a point need not be physically marked to be rendered a distinct object. For a point is nothing but a *place;* and a place is created such by the act of thinking it as an object. It is an *ens rationis.* Therefore, if anybody asks, Is there not a point on the path of the lowest particle of a pendulum where the pendulum is in its lowest place, I answer, yes. The act of thinking definitely of that point gives it all the reality it is in its nature to have.

18

['Continuity' in The Century Dictionary*]*

[(Whitney 1889), (Peirce 1888(?)–1914(?))] Peirce's definition of 'continuity' for the *Century Dictionary* was written around 1884; its significance for his own theory of continuity is mainly negative. In his dictionary entry he declares for Cantor's definition, of which he gives a largely, but not completely, accurate account. Take particular note of Peirce's use of time as the prime example of a continuum, and his use of Kant and Aristotle as reference points for the location of his own view; these are recurring themes in his explorations of continuity.

Peirce's critical reflections on the foregoing definition were handwritten in his copy of the *Century Dictionary;* they nicely encapsulate the development of his thinking on the subject.[1] The first note is a partially successful correction of an error in his earlier exposition of Cantor's definition, and the third reports developments which will be covered in detail in selection 19. In the second note, Peirce articulates a basic adequacy criterion for any definition of continuity, which must ensure that there are no gaps in a continuous line: no places where there could be, but is not, a point. Peirce suggests that this involves the idea of passing from one side of the line to the other; if the line is continuous, it should not be possible to do this without passing through a point of the line.

Peirce's criticism of Cantor's definition in that note, "as involving a vague reference to *all* the points," is unfair, strictly speaking: Cantor had in fact been quite specific about that (see note 7). But this note nonetheless provides a vital clue to *Peirce's* angle of approach to the problem of continuity: despite its unfairness, "it looms as the most important of the three in the development of Peirce's concept of continuity" (Noble 1989, 150). In that note he demands a completeness condition, which will make precise the absence of gaps in a true continuum. This is roughly Dedekind's approach in his classic treatise on continuity: having distinguished the rational points on the line, he asks how to fill in all the gaps they manifestly leave (Dedekind 1872, 6–21); his cuts are supposed to provide a systematic method of supplying all the missing points. Peirce proceeds similarly in selection 19, as summarized in his third note here. But while he always saw unbrokenness, the absence of gaps, as essential to true continuity, Peirce came to think that unbrokenness was not a matter of finding a sufficiently rich point set, big enough to fill up the line. From around 1896 on, the denial that *any* point set can fill a truly continuous line becomes a fundamental tenet of Peirce's theory of continuity.

The last two notes on the *Century Dictionary* definition take this stance, and explore its ramifications and difficulties. A major ramification is that points can no longer be regarded as parts of lines, as they are for Cantor and Dedekind. In both of these notes Peirce endorses, as the guiding principle of a proper understanding of continuity, Kant's definition of a continuum "as that all of whose parts have parts of the same kind." After clearing up some of Kant's, and his own, confusions about the meaning of this definition, Peirce tries to explain how points *do* relate to the line, if not as parts of it. It is in attempting to resolve this difficulty that Peirce resorts to the view of points as possibilia that we have already seen at work in selection 17. In the fourth note this view is seen to involve considerations of multitude (cardinality), whereas the fifth speaks only of the connection of the parts of a true continuum, and makes no direct mention of multitude. This difference of emphasis in these last two notes, marked by a shift from cardinality to (topological) connection as the leading idea in the theory of continuity, corresponds to the apparent doctrinal break between selections 26–29 and those that go before.

THE *CENTURY DICTIONARY* DEFINITION OF 'CONTINUITY' (CA. 1884)

1. Uninterrupted connection of parts in space or time; uninterruptedness.

2. In *math.* and *philos.*, a connection of points (or other elements) as intimate as that of the instants or points of an interval of time: thus, the *continuity* of space consists in this, that a point can move from any one position to any other so that at each instant it shall have a definite and distinct position in space. This statement is not, however, a proper definition of *continuity,* but only an exemplification drawn from time. The old definitions—the fact that adjacent parts have their limits in common (Aristotle),[2] infinite divisibility (Kant),[3] the fact that between any two points there is a third (which is true of the system of rational numbers)—are inadequate. The less unsatisfactory definition is that of G. Cantor, that continuity is the *perfect concatenation* of a system of points—words which must be understood in special senses.[4] Cantor calls a system of points *concatenated* when any two of them being given, and also any finite distance, however small, it is always possible to find a finite number of other points of the system through which

by successive steps, each less than the given distance, it would be possible to proceed from one of the given points to the other. He terms a system of points *perfect* when, whatever point not belonging to the system be given, it is possible to find a finite distance so small that there are not an infinite number of points of the system within that distance of the given point. As examples of a concatenated system not perfect, Cantor gives the rational and also the irrational numbers in any interval. As an example of a perfect system not concatenated, he gives all the numbers whose expression in decimals, however far carried out, would contain no figures except 0 and 9.[5]

NOTE 1 (CA. 1888–1892)

I here slightly modify Cantor's definition of a perfect system. Namely, he defines it as such that it contains every point in the neighborhood of an infinity of points and no other. But the latter is a character of a concatenated system; hence I omit it as a character of a perfect system.[6]

NOTE 2 (CA. 1892)

Cantor's definition of continuity is unsatisfactory as involving a vague reference to *all* the points, and one knows not what that may mean.[7] It seems to me to point to this: that it is impossible to get the idea of continuity without two dimensions. An oval line is continuous, because it is impossible to pass from the inside to the outside without passing a point of the curve.

NOTE 3 (CA. 1893)

Subsequent to writing the above I made a new definition, according to which continuity consists in *Kanticity* and *Aristotelicity.* The Kanticity is having a point between any two points. The Aristotelicity is having every point that is a limit to an infinite series of points that belong to the system.

Note 4 (18 September 1903)

But further study of the subject has proved that this definition is wrong. It involves a misunderstanding of Kant's definition which *he himself* likewise fell into. Namely he defines a continuum as that all of whose parts have parts of the same kind. He himself, and I after him, understood that to mean infinite divisibility, which plainly is not what constitutes continuity since the series of rational fractional values is infinitely divisible but is not by anybody regarded as continuous. Kant's real definition implies that a continuous line contains no points. Now if we are to accept the common-sense idea of continuity (after correcting its vagueness and fixing it to mean something) we must either say that a continuous line contains no points or we must say that the principle of excluded middle does not hold of those points. The principle of excluded middle only applies to an individual (for it is not true that 'Any man is wise' nor that 'Any man is not wise'). But places being mere possibles without actual existence are not individuals. Hence a point or indivisible place really does not exist unless there actually be something there to mark it, which, if there is, interrupts the continuity. I, therefore, think that Kant's definition correctly defines the common-sense idea, although there are great difficulties with it. I certainly think that on any line whatever, on the common-sense idea, there is room for any multitude of points however great. If so, the *analytical continuity* of the theory of functions which implies there is but a single point for each distance from the origin defined by a quantity expressible to indefinitely close approximation by a decimal carried out to an indefinitely great number of places, is certainly not the continuity of common sense since the whole multitude of such quantities is only the first abnumeral multitude,[8] and there is an infinite series of higher grades.

Note 5 (ca. 1903–1904)

Continuity, continued. On the whole, therefore, I think we must say that continuity is the relation of the parts of an unbroken space or time. The precise definition is still in doubt; but Kant's definition that a continuum is that of which every part has itself parts of the same kind, seems to be correct. This must not be confounded (as Kant himself confounded it) with infinite divisibility, but implies that a line, for example, contains no points until the continuity is broken by marking the points. In accordance with this, it seems necessary to say that a continuum, where it *is* continuous and unbroken, contains no definite parts; that its parts are created in the act of defining them and the precise definition of them breaks the continuity. In the calculus and theory of functions it is assumed that between any two rational points (or points at distances along the line expressed by rational fractions) there are rational points and that further for every convergent series of such fractions (such as 3.1, 3.14, 3.141, 3.1415, 3.14159 etc.) there is just one limiting point; and such a collection of points is called *continuous*. But this does not seem to be the common sense idea of continuity. It is only a collection of independent points. Breaking grains of sand more and more will only make the sand more broken. It will not weld the grains into unbroken continuity.

19

The Law of Mind

[*Peirce 1892*] In the early 1890s Peirce began to insist on the central importance of continuity in his philosophical system. This selection is taken from one of his first attempts to explain at length what continuity is and why it matters philosophically. The title of the piece highlights the role of continuity in relating ideas in the mind, but this turns out to involve the continuity of time; and in sections of the paper that are not included here Peirce applies continuity to general ideas, logical inference, personality and communication.

The "one law of mind" requires, as does the more traditional association of ideas, that ideas should affect one another. Peirce argues that they cannot do this unless they are present together in consciousness: how could an idea *out of* consciousness have any effect on an idea *in* it? Peirce's answer is that consciousness must take in, not an unextended temporal instant, but rather an interval of time, so that ideas in the earlier portions of the interval can affect later ones. (Indeed, if consciousness did not cover an interval, we could not even conceive, let alone know anything, of time itself.) Now if an idea can be affected by ideas whose date lies some finite temporal distance—say one second—in the past, then a present idea can be affected by an idea one second old, which can be affected by another two seconds old, and so on. The absurd consequence is that a present idea can be affected by all the past ideas of its containing consciousness. This *reductio,* together with the earlier argument that consciousness does cover a temporal interval, leads Peirce to the conclusion that it must cover an infinitesimal interval: an interval smaller than any finite span of time. Time, then, contains infinitesimal intervals. If time is itself a continuum, then it follows that at least one continuum contains infinitesimal intervals; Peirce's point, though he states it more clearly elsewhere than he does here, appears to be rather that *all* continua do; for he opens the following section on infinity by asserting the utility—indeed the indispensability—of infinitesimals for an adequate development of the calculus. Nor does he argue explicitly that time is continuous, though he does remark in passing that consciousness must be continuous through the infinitesimal interval in which ideas can affect one another directly.

Here, as in selection 18, Cantor is Peirce's chief inspiration and foil. He refers more widely to Cantor's early papers here than he did there (where he used only one section of one paper, the *Grundlagen* of 1883), and quotes a number of important Cantorian results about infinity and continuity. At the same time he expounds his own definitions of finiteness and countable infinity. Each of these has its own distinctive inference: the Syllogism of Transposed Quantity for finite collections, and the Fermatian Inference (mathematical induction) for countably infinite ones. What Cantor brings to the table is what Peirce calls *innumerable* (as we would say, uncountably infinite) collections. Cantor famously proved that the multitude of the real numbers is strictly greater than that of the natural numbers. By 1892 he had shown that there is in fact no greatest infinite number; Peirce, however, seems not to have been aware of that when he wrote "The Law of Mind," and asserts that there are only two infinite multitudes. He endorses Cantor's theory of infinity without significant qualification, both as the outstanding source of information (his own contributions aside) about the infinite, and as proof that infinity is a legitimate subject of mathematical study.

Peirce is more reserved about Cantor's definition of continuity, however. He lodges three complaints against it: (1) that it is, unlike the concept it is defining, metrical; (2) that it involves an illegitimate "definition by negation"; and (3) that it is insufficiently perspicuous. The second and most important of these criticisms has already shown up in the second of the five critical notes in selection 18.

Kant and Aristotle, who figured in selection 18 as mistaken precursors to Cantor, are honored here as possessors (and namesakes) of partial truths about continuity, which add up to an adequate definition. Kant recognized what is now known as the density of the continuum: there is a point between any two distinct points in it. Kanticity (as Peirce calls it) is only half the story, however; for it still allows there to be gaps in the continuum. Peirce finds the missing completeness principle by noting what happens when there *is* a gap; the resulting principle is named after Aristotle, whose conception of continuity involved an important special case of it. Aristotelicity requires that a bounded monotone sequence of points in the continuum have a supremum (least upper bound). Peirce's definition bears a strong resemblance to Dedekind's (1872, 772–775) definition in terms of cuts, which is itself equivalent to Cantor's (Cantor 1872, 92–96) in terms of Cauchy sequences of rational numbers.[1] Dedekind's continuum, however, satisfies a stronger completeness principle, which guarantees an upper bound for every bounded *set*—not just every bounded sequence—of points.

So Peirce's definition is not quite equivalent to Dedekind's, and he carefully refrains from asserting its equivalence to Cantor's, leaving open the possibility that Cantor's definition might be satisfied by a series that is not continuous as Peirce defines the term. Bear in mind that that when Peirce says this he is not comparing his definition with Cantor's 1872 definition just mentioned, but with the 1883 definition expounded in his entry in the *Century Dictionary*. He seems to take the latter, and not the former, to be Cantor's definition of the real numbers; otherwise it would be quite bizarre for him to express doubts, as he does here, about whether every real number is the limit of a Cauchy sequence of rationals! Similarly, when Peirce says that "there are as many points on a line or in an interval of time as there are of real numbers in all," he is (perhaps without realizing it) not giving 'real number' the meaning that has come down to us from Cantor and Dedekind. He would soon come to understand them better, and to separate his conception of continuity more sharply from theirs. But at this point he is still feeling his way.

Given these errors in Peirce's interpretation of Cantor, it is not all that surprising that he failed to recognize how radically he and Cantor disagreed over infinitesimals. Cantor was violently opposed to what he called the "cholera bacillus" of the infinitely small, and saw it as a virtue of his theory of continuity that it ruled out infinitesimals.[2] Peirce purports to show, with a dodgy argument about the "infinitieth place" in the decimal representation of an irrational number, that his own definition involves the infinitely small. It is easy to see that this has to be wrong; for Aristotelicity already rules out infinitesimals. (A monotone infinite sequence of infinitesimal steps, beginning at 0, would be bounded above by any finite number; but it would have no least upper bound, since the sum of two infinitesimals is itself infinitesimal, and there is no smallest finite positive number.)

Taking all this into account, it is not hard to see why, in selection 26, Peirce will refer to this analysis of continuity as his "blundering treatment" of the subject. It is mistaken about the content and implications, not just of Cantor's theory of continuity, but of Peirce's theory as well. But the significance of "The Law of Mind" is not altogether negative. The paper as a whole is of course a milestone in the evolution of Peirce's later philosophy, and its philosophical motivations would outlast the "blundering," and would also survive more pronounced and self-conscious departures from Cantor.

There are portents of those departures in "The Law of Mind." When Peirce touches on what he calls "endlessly infinite collections," he remarks that the "single individuals of such a collection could not, however, be designated, even approximately," thereby foreshadowing the "supermultitudinous" conception of the continuum that informs selections 21–25. The same can be said of the application of the definition he puts forth here, at the end of this selection, to the apparently paradoxical boundary phenomena that gave such a strong and enduring impetus to his inquiries into continuity.[3] The crux is the relationship between phenomena—color, velocity—at points and over continuous expanses. The residual tensions in the definition emerge when he says of the same surface both that (a) each of its points is either red or blue, and also that (b) the boundary between the two colors is half red, half blue. This strains at the standard view that the boundary is made up of points; but the strain has not quite reached the breaking point.

WHAT THE LAW IS.

Logical analysis applied to mental phenomena shows that there is but one law of mind, namely, that ideas tend to spread continuously and to affect certain others which stand to them in a peculiar relation of affectibility. In this spreading they lose intensity, and especially the power of affecting others, but gain generality and become welded with other ideas.

I set down this formula at the beginning, for convenience; and now proceed to comment upon it.

INDIVIDUALITY OF IDEAS.

We are accustomed to speak of ideas as reproduced, as passed from mind to mind, as similar or dissimilar to one another, and, in short, as if they were substantial things; nor can any reasonable objection be raised to such expressions. But taking the word "idea" in the sense of an event in an individual consciousness, it is clear that an idea once past is gone forever, and any supposed recurrence of it is another idea. These two ideas are not present in the same state of consciousness, and therefore cannot possibly be compared. To say, therefore, that they are similar can only mean that an occult power from the depths of the soul forces us to connect them in our thoughts after they are both no more. We may note, here, in passing that of the two generally recognised principles of association, contiguity and similarity, the former is a connection due to a power without, the latter a connection due to a power within.

But what can it mean to say that ideas wholly past are thought of at all, any longer? They are utterly unknowable. What distinct meaning can attach to saying that an idea in the past in any way affects an idea in the future, from which it is completely detached? A phrase between the assertion and the denial of which there can in no case be any sensible difference is mere gibberish.

I will not dwell further upon this point, because it is a commonplace of philosophy.

CONTINUITY OF IDEAS.

We have here before us a question of difficulty, analogous to the question of nominalism and realism. But when once it has been clearly formulated, logic leaves room for one answer only. How can a past idea be present? Can it be present vicariously? To a certain extent, perhaps; but not merely so; for then the question would arise how the past idea can be related to its vicarious representation. The relation, being between ideas, can only exist in some consciousness: now that past idea was in no consciousness but that past consciousness that alone contained it; and that did not embrace the vicarious idea.

Some minds will here jump to the conclusion that a past idea cannot in any sense be present. But that is hasty and illogical. How extravagant, too, to pronounce our whole knowledge of the past to be mere delusion! Yet it would seem that the past is as completely beyond the bounds of possible experience as a Kantian thing-in-itself.

How can a past idea be present? Not vicariously. Then, only by direct perception. In other words, to be present, it must be *ipso facto* present. That is, it cannot be wholly past; it can only be going, infinitesimally past, less past than any assignable past date. We are thus brought to the conclusion that the present is connected with the past by a series of real infinitesimal steps.

It has already been suggested by psychologists that consciousness necessarily embraces an interval of time. But if a finite time be meant, the opinion is not tenable. If the sensation that precedes the present by half a second were still immediately before me, then, on the same principle the sensation preceding that would be immediately present, and so on *ad infinitum*. Now, since there is a time, say a year, at the end of which an idea is no longer *ipso facto* present, it follows that this is true of any finite interval, however short.

But yet consciousness must essentially cover an interval of time; for if it did not, we could gain no knowledge of time, and not merely no veracious cognition of it, but no conception whatever. We are, therefore, forced to say that we are immediately conscious through an infinitesimal interval of time.

This is all that is requisite. For, in this infinitesimal interval, not only is consciousness continuous in a subjective sense, that is, considered as a subject or substance having the attribute of duration; but also, because it is immediate consciousness, its object is *ipso facto* continuous. In fact, this infinitesimally spread-out consciousness is a direct feeling of its contents as spread out. This will be further elucidated below. In an infinitesimal interval we directly perceive the temporal sequence of its beginning, middle, and end,—not, of course, in the way of recognition, for recognition is only of the past, but in the way of immediate feeling. Now upon this interval follows another, whose beginning is the middle of the former, and whose middle is the end of the former. Here, we have an immediate perception of the temporal sequence of its beginning, middle, and end, or say of the second, third, and fourth instants. From these two immediate perceptions, we gain a mediate, or inferential, perception of the relation of all four instants. This mediate perception is objectively, or as to the object represented, spread over the four instants; but subjectively, or as itself the subject of duration, it is completely embraced in the second moment. (The reader will observe that I use the word *instant* to mean a point of time, and *moment* to mean an infinitesimal duration.) If it is objected that, upon the theory proposed, we must have more than a mediate perception of the succession of the four instants, I grant it; for the sum of the two infinitesimal intervals is itself infinitesimal, so that it is immediately perceived. It is immediately perceived in the whole interval, but only mediately perceived in the last two thirds of the interval. Now, let there be an indefinite succession of these inferential acts of comparative perception; and it is plain that the last moment will contain objectively the whole series. Let there be, not merely an indefinite succession, but a continuous flow of inference through a finite time; and the result will be a mediate objective consciousness of the whole time in the last moment. In this last moment, the whole series will be recognised, or known as known before, except only the last moment, which of course will be absolutely unrecognisable to itself. Indeed, even this last moment will be recognised like the rest, or, at least be just beginning to be so. There is a little *elenchus,* or appearance of contradiction, here, which the ordinary logic of reflection quite suffices to resolve.

INFINITY AND CONTINUITY, IN GENERAL.

Most of the mathematicians who during the last two generations have treated the differential calculus have been of the opinion that an infinitesimal quantity is an absurdity; although, with their habitual caution, they have often added "or, at any rate, the conception of an infinitesimal is so difficult, that we practically cannot reason about it with confidence and security." Accordingly, the doctrine of limits has been invented to evade the difficulty, or, as some say, to explain the signification of the word "infinitesimal." This doctrine, in one form or another, is taught in all the text-books, though in some of them only as an alternative view of the matter; it answers well enough the purposes of calculation, though even in that application it has its difficulties.

The illumination of the subject by a strict notation for the logic of relatives had shown me clearly and evidently that the idea of an infinitesimal involves no contradiction, before I became acquainted with the writings of Dr. Georg Cantor (though many of these had already appeared in the *Mathematische Annalen* and in *Borchardt's Journal,* if not yet in the *Acta Mathematica,* all mathematical journals of the first distinction), in which the same view is defended with extraordinary genius and penetrating logic.[4]

The prevalent opinion is that finite numbers are the only ones that we can reason about, at least, in any ordinary mode of reasoning, or, as some authors express it, they are the only numbers that can be reasoned about mathematically. But this is an irrational prejudice. I long ago showed that finite collections are distinguished from infinite ones only by one circumstance and its consequences, namely that to them is applicable a peculiar and unusual mode of reasoning called by its discoverer, De Morgan, the "syllogism of transposed quantity."[5]

Balzac, in the introduction of his *Physiologie du mariage,* remarks that every young Frenchman boasts of having seduced some Frenchwoman. Now, as a woman can only be seduced once, and there are no more Frenchwomen than Frenchmen, it follows, if these boasts are true, that no French women escape seduction. If their number be finite, the reasoning holds. But since the population is continually increasing, and the seduced are on the average younger than the seducers, the conclusion need not be true. In like manner, De Morgan, as an actuary, might have argued that if an insurance company pays to its insured on an average more than they have ever paid it, including interest, it must lose money. But every modern actuary would see a fallacy in that, since the business is continually on the increase. But should war, or other cataclysm, cause the class of insured to be a finite one, the conclusion would turn out painfully correct, after all. The above two reasonings are examples of the syllogism of transposed quantity.

The proposition that finite and infinite collections are distinguished by the applicability to the former of the syllogism of transposed quantity ought to be regarded as the basal one of scientific arithmetic.

If a person does not know how to reason logically, and I must say that a great many fairly good mathematicians,—yea distinguished ones,—fall under this category, but simply uses a rule of thumb in blindly drawing inferences like other inferences that have turned out well, he will, of course, be continually falling into error about infinite numbers. The truth is such people do not reason, at all. But for the few who do reason, reasoning about infinite numbers is easier than about finite numbers, because the complicated syllogism of transposed quantity is not called for. For example, that the whole is greater than its part is not an axiom, as that eminently bad reasoner, Euclid, made it to be. It is a theorem readily proved by means of a syllogism of transposed quantity, but not otherwise. Of finite collections it is true, of infinite collections false. Thus, a part of the whole numbers are even numbers. Yet the even numbers are no fewer than all the numbers; an evident proposition since if every number in the whole series of whole numbers be doubled, the result will be the series of even numbers.

> 1, 2, 3, 4, 5, 6, etc.
> 2, 4, 6, 8, 10, 12, etc.

So for every number there is a distinct even number. In fact, there are as many distinct doubles of numbers as there are of distinct numbers. But the doubles of numbers are all even numbers.

In truth, of infinite collections there are but two grades of magnitude, the *endless* and the *innumerable*.[6] Just as a finite collection is distinguished from an infinite one by the applicability to it of a special mode of reasoning, the syllogism of transposed quantity, so, as I showed in the paper last referred to,[7] a numerable collection is distinguished from an innumerable one by the applicability to it of a certain mode of reasoning, the Fermatian inference, or, as it is sometimes improperly termed, "mathematical induction."[8] . . .

Such reasoning holds good of any collection of objects capable of being ranged in a series which though it may be endless, can be numbered so that each member of it receives a definite integral number. For instance, all the whole numbers constitute such a numerable collection. Again, all numbers resulting from operating according to any definite rule with any finite number of whole numbers form such a collection.[9] For they may be arranged in a series thus. Let F be the symbol of operation. First operate on 1, giving $F(1)$. Then, operate on a second 1, giving $F(1,1)$. Next, introduce 2, giving 3rd, $F(2)$; 4th, $F(2,1)$; 5th, $F(1,2)$; 6th, $F(2,2)$. Next use a third variable giving 7th, $F(1,1,1)$; 8th, $F(2,1,1)$; 9th, $F(1,2,1)$; 10th, $F(2,2,1)$; 11th, $F(1,1,2)$; 12th, $F(2,1,2)$; 13th, $F(1,2,2)$; 14th, $F(2,2,2)$. Next introduce 3, and so on, alternately introducing new variables and new figures; and in this way it is plain that every arrangement of integral values of the variables will receive a numbered place in the series.[*]

The class of endless but numerable collections (so called because they can be so ranged that to each one corresponds a distinct whole number) is very large. But there are collections which are certainly innumerable. Such is the collection of all numbers to which endless series of decimals are capable of approximating. It has been recognised since the time of Euclid that certain numbers are surd or incommensurable, and are not exactly expressible by any finite series of decimals, nor by a circulating decimal. Such is the ratio of the circumference of a circle to its diameter, which we know is nearly 3.1415926. The calculation of this number has been carried to over 700 figures without the slightest appearance of regularity in their sequence. The demonstrations that this and many other numbers are incommensurable are perfect. That the entire collection of incommensurable numbers is innumerable has been clearly proved by Cantor. I omit the demonstration; but it is easy to see that to discriminate one from some other would, in general, require the use of an endless series of numbers. Now if they cannot be exactly expressed and discriminated, clearly they cannot be ranged in a linear series.[10]

It is evident that there are as many points on a line or in an interval of time as there are of real numbers in all. These are, therefore, innumerable collections. Many mathematicians have incautiously assumed that the points on a surface or in a solid are more than those on a line. But this has been refuted by Cantor. Indeed, it is obvious that for every set of values of coördinates there is a single distinct number. Suppose, for instance, the values of the coördinates all lie between 0 and +1. Then if we compose a number by putting in the first decimal place the first figure of the first coördinate, in the second the first figure of the second coördinate, and so on, and when the first figures are all dealt out go on to the second figures in like manner, it is plain

---

[*]. This proposition is substantially the same as a theorem of Cantor, though it is enunciated in a much more general form.

that the values of the coördinates can be read off from the single resulting number, so that a triad or tetrad of numbers, each having innumerable values, has no more values than a single incommensurable number.[11]

Were the number of dimensions infinite, this would fail; and the collection of infinite sets of numbers, having each innumerable variations, might, therefore, be greater than the simple innumerable collection, and might be called *endlessly infinite.* The single individuals of such a collection could not, however, be designated, even approximately, so that this is indeed a magnitude concerning which it would be possible to reason only in the most general way, if at all.

Although there are but two grades of magnitudes of infinite collections, yet when certain conditions are imposed upon the order in which individuals are taken, distinctions of magnitude arise from that cause. Thus, if a simply endless series be doubled by separating each unit into two parts, the successive first parts and also the second parts being taken in the same order as the units from which they are derived, this double endless series will, so long as it is taken in that order, appear as twice as large as the original series. In like manner the product of two innumerable collections, that is, the collection of possible pairs composed of one individual of each, if the order of continuity is to be maintained, is, by virtue of that order, infinitely greater than either of the component collections.[12]

We now come to the difficult question, What is continuity? Kant confounds it with infinite divisibility, saying that the essential character of a continuous series is that between any two members of it a third can always be found. This is an analysis beautifully clear and definite; but unfortunately, it breaks down under the first test. For according to this, the entire series of rational fractions arranged in the order of their magnitude, would be an infinite series, although the rational fractions are numerable, while the points of a line are innumerable. Nay, worse yet, if from that series of fractions any two with all that lie between them be excised, and any number of such finite gaps be made, Kant's definition is still true of the series, though it has lost all appearance of continuity.

Cantor defines a continuous series as one which is *concatenated* and *perfect.* By a concatenated series, he means such a one that if any two points are given in it, and any finite distance, however small, it is possible to proceed from the first point to the second through a succession of points of the series each at a distance from the preceding one less than the given distance. This is true of the series of rational fractions ranged in the order of their magnitude. By a perfect series, he means one which contains every point such that there is no distance so small that this point has not an infinity of points of the series within that distance of it. This is true of the series of numbers between 0 and 1 capable of being expressed by decimals in which only the digits 0 and 1 occur.[13]

It must be granted that Cantor's definition includes every series that is continuous; nor can it be objected that it includes any important or indubitable case of a series not continuous. Nevertheless, it has some serious defects. In the first place, it turns upon metrical considerations; while the distinction between a continuous and a discontinuous series is manifestly non-metrical. In the next place, a perfect series is defined as one containing "every point" of a certain description. But no positive idea is conveyed of what all the points are: that is definition by negation, and cannot be admitted. If that sort of thing were allowed, it would be very easy to say, at once, that the continuous linear series of points is one which contains every point of the line between its extremities. Finally, Cantor's definition does not convey a distinct notion of what the components of the conception of continuity are. It ingeniously wraps up its properties in two separate parcels, but does not display them to our intelligence.

Kant's definition expresses one simple property of a continuum; but it allows of gaps in the series. To mend the definition, it is only necessary to notice how these gaps can occur. Let us suppose, then, a linear series of points extending from a point, *A*, to a point, *B*, having a gap from *B* to a third point, *C*, and thence extending to a final limit, *D*; and let us suppose this series conforms to Kant's definition. Then, of the two points, *B* and *C*, one or both must be excluded from the series; for otherwise, by the definition, there would be points between them. That is, if the series contains *C*, though it contains all the points up to *B*, it cannot contain *B*. What is required, therefore, is to state in non-metrical terms that if a series of points up to a limit is included in a continuum the limit is included. It may be remarked that this is the property of a continuum to which Aristotle's attention seems to have been directed when he defines a continuum as something whose parts have a common limit. The property may be exactly stated as follows: If a linear series of points is continuous between two points, *A* and *D*, and if an endless series of points be taken, the first of them between *A* and *D* and each of the others between the last preceding one and *D*, then there is a point of the continuous series between all that endless series of points and *D*, and such that every other point of which this is true lies between this point and *D*. For example, take any number between 0 and 1, as 0.1; then, any number between 0.1 and 1, as 0.11; then any number between 0.11 and 1, as 0.111;

and so on, without end. Then, because the series of real numbers between 0 and 1 is continuous, there must be a *least* real number, greater than every number of that endless series. This property, which may be called the Aristotelicity of the series, together with Kant's property, or its Kanticity, completes the definition of a continuous series.

The property of Aristotelicity may be roughly stated thus: a continuum contains the end point belonging to every endless series of points which it contains. An obvious corollary is that every continuum contains its limits. But in using this principle it is necessary to observe that a series may be continuous except in this, that it omits one or both of the limits.

Our ideas will find expression more conveniently if, instead of points upon a line, we speak of real numbers. Every real number is, in one sense, the limit of a series, for it can be indefinitely approximated to. Whether every real number is a limit of a *regular* series may perhaps be open to doubt.[14] But the series referred to in the definition of Aristotelicity must be understood as including all series whether regular or not. Consequently, it is implied that between any two points an innumerable series of points can be taken.

Every number whose expression in decimals requires but a finite number of places of decimals is commensurable. Therefore, incommensurable numbers suppose an infinitieth place of decimals. The word infinitesimal is simply the Latin form of infinitieth; that is, it is an ordinal formed from *infinitum,* as centesimal from *centum.* Thus, continuity supposes infinitesimal quantities. There is nothing contradictory about the idea of such quantities. In adding and multiplying them the continuity must not be broken up, and consequently they are precisely like any other quantities, except that neither the syllogism of transposed quantity, nor the Fermatian inference applies to them.

If A is a finite quantity and i an infinitesimal, then in a certain sense we may write $A+i = A$. That is to say, this is so for all purposes of measurement. But this principle must not be applied except to get rid of all the terms in the highest order of infinitesimals present. As a mathematician, I prefer the method of infinitesimals to that of limits, as far easier and less infested with snares. Indeed, the latter, as stated in some books, involves propositions that are false; but this is not the case with the forms of the method used by Cauchy, Duhamel, and others. As they understand the doctrine of limits, it involves the notion of continuity, and therefore contains in another shape the very same ideas as the doctrine of infinitesimals.[15]

Let us now consider an aspect of the Aristotelical principle which is particularly important in philosophy. Suppose a surface to be part red and part blue; so that every point on it is either red or blue, and, of course, no part can be both red and blue. What, then, is the color of the boundary line between the red and the blue? The answer is that red or blue, to exist at all, must be spread over a surface; and the color of the surface is the color of the surface in the immediate neighborhood of the point. I purposely use a vague form of expression. Now, as the parts of the surface in the immediate neighborhood of any ordinary point upon a curved boundary are half of them red and half blue, it follows that the boundary is half red and half blue. In like manner, we find it necessary to hold that consciousness essentially occupies time; and what is present to the mind at any ordinary instant, is what is present during a moment in which that instant occurs. Thus, the present is half past and half to come. Again, the color of the parts of a surface at any finite distance from a point, has nothing to do with its color just at that point; and, in the parallel, the feeling at any finite interval from the present has nothing to do with the present feeling, except vicariously. Take another case: the velocity of a particle at any instant of time is its mean velocity during an infinitesimal instant in which that time is contained. Just so my immediate feeling is my feeling through an infinitesimal duration containing the present instant.

20

[Scientific Fallibilism]

[Peirce 1893d] In the summer of 1893 Peirce wrote a lecture on "Scientific Fallibilism," which has only recently been reconstructed (De Tienne 2001) from a number of scattered manuscripts; one of these, the source of this selection, was excerpted separately in volume 1 of the *Collected Papers* under the editor-supplied title "Fallibilism, Continuity and Evolution." The interconnections that Peirce draws among the three themes of that title have already been discussed at some length in the introduction to this book. The other main topic of this selection is the justification of *synechism,* which at this point Peirce takes to be the doctrine that all things are continuous. As in the "Law of Mind," Peirce argues (in a passage omitted here) that the continuity of time is essential to an adequate account of consciousness. He is less confident about the continuity of space and of "degrees of quality"; here he gives not one definitive argument but several less conclusive ones, "some positive and others only formal, yet not contemptible." The strongest of these arguments, in his view, is methodological: synechism makes possible explanations of phenomena (the influence of minds on one another, and physical action at a distance) that a nominalist would hold to be inexplicable. Thus synechism enables us to comply with a basic Peircean imperative: do not block the road of inquiry.

But in order really to see all there is in the doctrine of fallibilism, it is necessary to introduce the idea of continuity, or unbrokenness. This is the leading idea of the differential calculus and of all the useful branches of mathematics; it plays a great part in all scientific thought, and the greater the more scientific that thought is; and it is the master-key which adepts tell us unlocks the *arcana* of philosophy.

We all have some idea of continuity. Continuity is fluidity the merging of part into part. But to achieve a really distinct and adequate conception of it is a difficult task, which with all the aids possible must for the most acute and most logically trained intellect require days of severe thought. If I were to attempt to give you any logical conception of it, I should only make you dizzy to no purpose. I may say this however. I draw a line

Now the points on that line form a continuous series. If I take away any two points on that line however close together, other points there are lying between them. If that were not so, the series of points would not be continuous. It might be so, even if the series of points were not continuous.

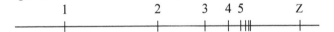

Now suppose I mark a point Z and another at the left of it 1. Suppose a point to be inserted between 1 and Z, and marked 2 and another between 2 and Z marked 3, and so on *ad infinitum.* Then there will be points at the right of all the numbered points,—for Z is such a point. And because the series of points on the line is continuous it follows that of all the points that are at the right of all the numbered points, there is one that is the furthest to the left, or closest to those numbered points.

You will readily see that the idea of continuity involves the idea of infinity. Now, the nominalists tell us that we cannot reason about infinity,—or that we cannot reason about it *mathematically.* Nothing can be more false. Nominalists cannot reason about infinity, because they do not reason logically about anything. Their reasoning consists of performing certain processes which they have found worked well,—without having any insight into the conditions of their working well. This is not logical reasoning. It naturally fails when infinity is involved; because they reason about infinity as if it were finite. But to a logical reasoner, reasoning about infinity is decidedly simpler than reasoning about finite quantity.

There is one property of a continuous expanse that I must mention, though I cannot venture to trouble you with the demonstration of it. It is that in a continuous expanse, say a continuous line, there are continuous lines infinitely short. In fact, the whole line is made up of such infinitesimal parts. The property of these infinitely small spaces is,—I regret the abstruseness of what I am going to say, but I cannot help it,—the property which distinguishes these infinitesimal distances is that a certain mode of reasoning which holds good of all finite quantities and of some that are not finite does not hold good of them. Namely, mark any point on the line A. Suppose that point to have any character; suppose, for instance, it is *blue.* Now suppose we lay down the rule that every point within an inch of a blue point shall be painted blue. Obviously, the consequence will be that the whole line will have to be blue. But this reasoning does not hold good of infinitesimal distances. After the point A has been painted blue, the rule that every point infinitesimally near to a blue point shall be painted blue will not necessarily result in making the whole blue.

Continuity involves infinity in the strictest sense, and infinity even in a less strict sense goes beyond the possibility of direct experience. . . .

No equally conclusive and direct reason for thinking that space and degrees of quality and other things are continuous is to be found as for believing time to be so. Yet, the reality of continuity once admitted reasons are there, divers reasons, some positive others only formal, yet not contemptible, for admitting the continuity of all things. I am making a bore of myself and won't bother with any full statement of these reasons but will just indicate the nature of a few of them. Among formal reasons, there are such as these, that it is easier to reason about continuity than about discontinuity, so that it is a convenient assumption. Also, in case of ignorance it is best to adopt the hypothesis which leaves open the greatest field of possibility; now a continuum is merely a discontinuous series with additional possibilities. Among positive reasons, we have the apparent analogy between time and space, between time and degree, and so on.

There are various other positive reasons; but the weightiest consideration appears to me to be this.

How can one mind act upon another mind? How can one particle of matter act upon another at a distance from it? The nominalists tell us this is an ultimate fact,—it cannot be explained. Now, if this were meant in [a] merely practical sense, if it were only meant that so much we know that one thing does act on another but that how it takes place we cannot very well tell, up to date, I should have nothing to say, except to applaud the moderation and good logic of the statement. But this is not what is meant; what is meant is that we come up bump against actions absolutely unintelligible and inexplicable, where human inquiries have to stop. Now that is a mere *theory,* and nothing can justify a theory except its explaining observed facts. It is a poor kind of theory which in place of performing this the sole legitimate function of a theory merely supposes the facts to be inexplicable. It is one of the peculiarities of nominalism that it is continually supposing things to be absolutely inexplicable. That blocks the road of inquiry. But if we adopt the theory of continuity, we escape this illogical situation. We may then say that one portion of mind acts upon another, because it is in a measure immediately present [to] that other; just as we suppose that the infinitesimally past is in a measure present. And in like manner we may suppose that one portion of matter acts upon another because it is in a measure in the same place.

If I were to attempt to describe to you in full all the scientific beauty and truth that I find in the principle of continuity, I may say in the simple language of Matilda in Engaged,[1] "the tomb would close over me e'er the entrancing topic were exhausted,"—but not before my audience was exhausted. So I will just drop it here. Only in doing so, let me call your attention to the natural affinity of this principle to the doctrine of fallibilism. The principle of continuity is the idea of fallibilism objectified. For fallibilism is the doctrine that our knowledge is never absolute but always swims as it were in [a] continuum of uncertainty and of indeterminacy. Now the doctrine of continuity is that *all things* so swim in continua.

The doctrine of continuity rests upon observed fact as we have seen. But what opens our eyes to the significance of that fact is fallibilism. The ordinary scientific infallibilist,—of which sect Büchner in his "Kraft und Stoff" affords a fine example,[2]—cannot accept *synechism*,—or the doctrine that all that exists is continuous,—because he is committed to discontinuity in regard to all those things which he fancies he has exactly ascertained, and especially in regard to that part of his knowledge which he fancies he has exactly ascertained to be *certain*. For where there is continuity, the exact ascertainment of real quantities is too obviously impossible. No sane man can dream that the ratio of the circumference to the diameter could be exactly ascertained by measurement. As to the quantities he has not yet exactly ascertained, the Büchnerite is naturally led to separate them into two distinct classes—those which may be ascertained hereafter[,] and there as before continuity must be excluded[,] and those absolutely unascertainable,—and these in their utter and everlasting severance from the other class present a new breach of continuity. Thus scientific infallibilism draws down a veil before the eyes which prevents the evidences of continuity from being discerned.

But as soon as a man is fully impressed with the fact that absolute exactitude never can be known, he naturally asks whether there are any facts to show that hard discrete exactitude really exists. That suggestion lifts the edge of that curtain and he begins to see the clear daylight shining in from behind it.

21

On Quantity [The Continuity of Time and Space]

[Peirce 1896(?)] Around the beginning of 1896 Peirce discovered a version of what has become known as Cantor's Theorem (for one of his proofs, see selection 22). A corollary of that theorem—that there is no greatest multitude—transformed his theory of the continuum. His quarrel with Cantor over "all the points" comes to take the form of a requirement of what he would eventually call *supermultitudinousness:* since a continuum must contain all the points it could contain, and since there is no greatest multitude, a continuum cannot be a collection of points with a definite multitude. Closely related to this is the idea that a continuum's points lack distinct identities. Both ideas are clearly present here, but Peirce does not yet use modality to connect them in a systematic way. The manuscript from which this selection is taken was probably written very soon after his discovery of Cantor's Theorem, and hence is quite exploratory in nature.

As the bracketed subtitle (taken from Peirce's title for this section of the manuscript) indicates, the primary topic of the selection is the continuity of space and time—especially of time, from whose continuity that of space is said to be derived. But Peirce begins with some pregnant remarks about logic. Echoing his ontological analyses of collections, he asks wherein the existence of a "general" (universal) consists. The traditional answer is that the general's existence depends upon that of its instances. Deny that, and generals turn out to be supermultitudinous continua.

The analysis of time opens with a quick restatement of the *reductio* from selection 19, and much of the argumentation here is cut from the same cloth. In what may be a backhanded acknowledgment of the strangeness of the supermultitudinous view, Peirce tries to show that it is forced upon us by our common sense ideas of space and time. Though this is his announced intention, the argument for the crucial premise that the flow of time is directly perceived looks more like a transcendental argument than a common sense one. Adapting a strategy already employed in selection 19, Peirce submits that we could not even imagine time if we did not directly perceive its flow. It follows, from the direct perception of temporal flow, that "something more than an instant is immediately present to consciousness." We can make sense of the "something more" if we suppose that a span of time has room for an arbitrarily large collection of instants, and that its instants merge into one another. These two features of time are then used to prove that time is super-multitudinous, and to analyze the boundary phenomena presented by change in time. The resulting model of the flow of time is clearly descended from "The Law of Mind"; immediate awareness of a temporal interval is now held to involve, along with awareness of the present instant, awareness of a denumerable (countable) succession of instants before and after.

This is not much of an explanation, if the denumerable succession is a countable infinity of discrete acts of awareness. But Peirce breaks off that discussion to give a new, three-part definition of the continuum. Continua are linearly ordered and infinitely capacious; these two parts are tolerably clear. But in the first of the three parts, where he tries to explain the part/whole relation in the continuum, Peirce has not yet broken the hold of the familiar view of a continuum as an "aggregate of . . . mutually exclusive subjects."

OF THE NATURE OF THE CONTINUITY OF TIME AND SPACE

Art. 21. Cantor, in effect, defines the continuity of a line as consisting in that line's containing *all* its points. This is a singular *circulus in definiendo,* since the very problem was to state how those points were related. But I should not have noticed it, were it not that the phrase seems to imply that the line contains as many points as it could contain. Now we have seen in the last section that there is no maximum grade of multitude. If, therefore, a line contains all the points there could be, these points must cease to form a multitude.

Logic, as it was conceived by Aristotle and as it was apprehended even by the subtlest realists of the middle ages and as its ideas are embodied in every development of syllogistic, rests as upon bed-rock upon the principle that a "general" exists only in so far as it inheres in individuals, which are the "first substances," having absolute, independent, and ultimate existence. Many philosophers have denied this in metaphysics; but they have never shown what would be left of formal logic after the havoc that denial would bring into that field. In the English logic of this century, generals appear as "class-names"; and a class is a multitude, or collection, of individual things, each having its distinct, independent, and prior existence. Such a class cannot have as many individuals as there could be; because "as many as there could be" is not a possible grade of multitude; and the result of insisting upon that would inevitably be that the individuals would be sunk to a potential being, and would no longer be unconditionally and *per se* there. The discovery of such a state of things would be an earthquake in logic, leveling its whole fabric; and it would be incumbent upon the philosopher who should accept it to begin at the very beginning and build up the elementary rules of reasoning anew.

We must either hold that there are not as many points upon a line as there might be, or else we must say that points are in some sense fictions which are freely made up when and where they are wanted.

As far as points upon a spatial line go, no doubt a large party would be quite disposed to regard them so. But the continuity of space seems unquestionably to be derived from the continuity of time; and common sense would find very grave difficulty in admitting that the smallest portion of past or future time was immediately present; and if not, the present instant would seem to be the most indubitable and independent reality in all our knowledge.

I consider that it is pertinent to the present investigation, first, to analyze the nature of the continuity of space, and especially of time, as logically involved in the common-sense ideas of those continua, and second, to consider what the evidence is that objective time and space possess such continuity. For although mathematics has nothing to do with positive truth, yet its hypotheses are suggested by experience, and any theory for which there may be even imperfect evidence ought to be erected into a mathematical hypothesis, provided it be of such a nature that a great body of deductions can be drawn from it. In short, though this part of the inquiry can only shed a sidelight on the main question, which pertains to the infinitesimals of the calculus, yet that illumination may be strong, and in my opinion will be so.

Art. 22. According to natural common-sense, only the state at a single instant, the present instant, is ever immediately present to consciousness, and yet we are conscious of the flow of time, we imagine events as in time, and we have a real memory, not merely of states, but also of motions. Our present task is, not to criticise this idea of time, but to endeavour to gain a distinct comprehension of its elements and of how they are related to one another. Imagine a series of instantaneous photographs to be taken. Then, no matter how closely they follow one another, there is no more motion visible in any one of them than if they were taken at intervals of centuries. This is the common-sense idea of that which is immediately present to the mind.

Opinions will differ as to whether common-sense holds that the flow of time is directly perceived or not. Let us first suppose that it does not. Then, according to common-sense, we can come to believe that events happen in time only by a sort of vaticination, the idea springing up in the mind without any reason. For reasoning must be conscious that it is reasoning, or it ceases to be reasoning. Now reasoning is essentially a process. Consequently, we cannot reason without having already the idea of time. A greater difficulty is that an instantaneous photograph, though it may contain a symbol of time, or even an indication that time exists, can certainly not contain a true likeness of time. To imagine time, time is required. Hence, if we do not directly perceive the flow of time, we cannot imagine time. Yet the sense of time is something forced upon common-sense. So that, if common-sense denies that the flow [of] time is directly perceived, it is hopelessly entangled in contradictions and cannot be identified with any distinct and intelligible conception.

But to me it seems clear that our natural common-sense belief is that the flow of time *is* directly perceived. In that case, common-sense must hold that something more than an instant is immediately present to consciousness. As to what this "something more" is, several hypotheses might be made. But there are two propositions about time which, if they are acknowledged to be involved in the common-sense idea, determine the character of its continuity. One is, that there is in a sensible time room for any multitude, however great, of distinct instants. The other is, that the instants are so close togther as to merge into one another, so that they are not distinct from one another. I do not think that anybody fairly considering the matter can doubt that the natural idea of time common to all men supports both those judgments. The former seems to express what there is that is true in Cantor's statement that a line includes *all* the points possible. Its consonance with the common idea is further shown by the circumstance that it has occurred to nobody to object to the orders of infinitesimals of the calculus that there would not be room in time or space for so many distinct points as they create, although the vagueness of the multitude needed has been strongly felt. Nor has there ever been any doubt that surds and transcendentals of all real kinds could be conceived as measured off in space (whose continuity is recognized as precisely like that of time), although there have been doubts as to whether the variations of those quantities were adequate to representing all the points of space. On the whole, then, this proposition may be accepted as a dictum of common-sense. The other proposition, that the instants of time are so crowded as to merge into one another and lose their distinct existence, it seems to be involved in the conception of the "flow" of time. For this phrase likens time to a homogeneous fluid in which the "particles" are mere creations of the mind, made for convenience of calculators. Again, nobody has, as far as I know, ever suggested that two line[s] might cross one another without having a common point, although if their points were distinct from one another,—two multitudes in series order,—there is no reason why the points of one line might not slip through between those of the other. The very word *continuity* implies that the instants of time or the points of a line are everywhere welded together. This proposition may then, likewise, be accepted as true to the common idea.

According to that idea, then, the instants of a time are not a multitude. Each of the two propositions proves that. For, first, since any multitude whatever of instants exists among the instants of any given time, and since there is no maximum multitude, it follows that the instants of time do not in their totality form a multitude. In this sense, they may be said to be "more" than any multitude; that is, there is among them a multitude greater than any multitude which may be proposed. Second, [they] are not in themselves distinct from one another, as the units of any multitude are, even if they happen to be joined together.

Moreover, because the instants do not preserve their distinct identities, it follows that taking any proposition whatsoever, if that proposition is true of any instant then a later instant can be found such that it remains true to that instant, and in like manner an earlier instant can be found such that it has been true since that instant. Were a proposition to be false up to a certain instant and thereafter to be true, at that instant it would be both true and false. It does not follow that a proposition once true remains always true; it only follows that it remains true through a denumerable series of instants, which is a lapse of time inexpressibly less than any sensible or assignable time, if it can properly be called a lapse of time, at all, wanting as it does most of the characteristics of duration. A denumerable succession of instants may be called a *moment.*

Even infinitely longer than the moment will a proposition of the slightest latitude in respect to the change which is taking place, though this latitude be far too small to be detected, remain true.

In particular, if an instant be *immediately present,*—since this is a proposition concerning that instant, a denumerable succession of instants before and after it are fully present; and even infinitely longer will the proximity of instants be so close, in the case of the past with respect to the action of facts upon the mind, through sense, and in the case of the future with respect to the action of the mind upon facts, through volition, that they are practically present. Thus, temporal succession is immediately present to consciousness, according to the logical explication of the common-sense idea of time; and even temporal continuity is practically present; and so there is no difficulty in accounting for the suggestion of the idea of time, nor for our being persuaded of its truth.

We are now prepared to define a *continuum* after the exemplar of the common-sense idea of time. Namely, a *continuum* is whatever has the following properties:

1st, it is a whole composed of parts. We must define this relation. The parts are a logical aggregate of mutually exclusive subjects having a common predicate; and that aggregate regarded as a single object is the whole.

2nd, these parts form a series. That is, there is a relation, l, such that, taking any two of the parts, if these are not identical one of them is in the relation, l, to everything to which the other is in that relation and to something else besides.

3rd, taking any multitude whatever, a collection of those parts can be found whose multitude is greater than the given multitude. Consequently, the indivisible parts, that is, parts such that none is a collective aggregate of objects one of which is in the relation, l, to everything to which another is in that relation and to more besides,—are not distinct. That is to say, the relation l cannot be fully defined, so that in any attempted specification of it, l', any part which appears indivisible, becomes divisible into others, by means of a further specification, l''.

22

Detached Ideas Continued and the Dispute between Nominalists and Realists

[*Peirce 1898a*] Not quite two years after writing selection 21, Peirce is in a position to give a more systematic account of continuity. This excerpt from his Cambridge Conferences Lectures of 1898 contains a very elegant argument for his version of Cantor's Theorem: for any collection *C* of distinct elements, the collection *S(C)* of its subcollections is itself a collection of distinct elements and has a greater multitude than *C*'s; furthermore, if *C* has the infinite multitude *M* then *S(C)*'s multitude is the next infinite multitude after *M*. (Here Peirce assumes something like what set theorists call the Generalized Continuum Hypothesis; it is now known to be independent of the usual axioms for set theory.)[1] He describes a construction in which we take the union of a collection containing, for every multitude *M,* a collection of multitude *M:* call this union *U*. Peirce asserts that *U* and *S(U)* have the same multitude. (He omits the—fallacious—argument for this claim: see selection 24.) So if the elements of *U* were distinct from one another, it would violate his theorem. Therefore *U* is a collection so large that its elements are no longer distinct: it is a continuous collection. (It is immediate from the construction that *U* is supermultitudinous.)

Peirce gives a striking series of illustrations of the resulting conception of points. He concedes that a truly continuous curve is, in a sense, a collection of points, but their lack of distinct identities makes it possible for one point to become two and vice versa, and for a single point to "explode" into an infinite multitude of points. Unfortunately he simply asserts that this is a consistent picture—his auditors are asked to take his word for it that the theory has passed muster with the logic of relatives.

This theory of continuity has been widely commented upon, and there have been a number of attempts to reconstruct it within more recent and more rigorously developed mathematical frameworks. Hilary Putnam's highly influential reconstruction (Putnam 1992) draws on Robinson's nonstandard analysis, whose similarities to Peirce's conception had already been remarked on by Carolyn Eisele (1979e, p. 215; 1979f, pp. 246–248). Timothy Herron (1997, 620–623), Jérôme Havenel (2008, 111–112), and John Bell (1998, p. 4; 2005, pp. 295–296), on the other hand, see Peirce more as a precursor of smooth infinitesimal analysis, which comes out of category theory. Fernando Zalamea (forthcoming) relates Peirce's continuum to a rather dazzling array of category-theoretic developments; see also Zalamea (2001; 2003). Partial reconstructions are offered by Arnold Johanson (Johanson 2001), using his own "pointless topology," and Wayne Myrvold (1995, 535–537), working in Zermelo-Fraenkel set theory. The most technically detailed reconstruction to date is due to Philip Ehrlich (forthcoming), working in the setting of his theory of absolute continua, and making connections to the wider history of non-Archimedean mathematics laid out in Ehrlich (2006).

Much of the omitted material in this lecture deals with Peirce's contributions to logic; the discussion of continuity is itself occasioned by the question of "what it is in the logic of relatives which takes the logical position occupied in ordinary logic by 'generality,' or in medieval language by the *universal*." So Peirce is following up here on the tantalizing remarks on logic in selection 21, though he cannot be said to follow them all the way to the end. The original lecture also contains a useful exposition of Peirce's categories; it has been retained here as background for this selection and others in this volume.

When in 1866, Gentlemen, I had clearly ascertained that the three types of reasoning were Induction, Deduction, and Retroduction,[2] it seemed to me that I had come into possession of a pretty well-rounded system of Formal Logic. I had, it is true, a decided suspicion that there might be a logic of relations; but still I thought that the system I had already obtained ought to enable me to take the Kantian step of transferring the conceptions of logic to metaphysics. My formal logic was marked by triads in all its principal parts. There are three types of inference Induction, Deduction, and Retroduction each having three propositions and three terms. There are three types of logical forms, the term, the proposition, and the inference. Logic is itself a study of signs. Now a sign is a thing which represents a second thing to a third thing, the interpreting thought. There are three ways in which signs can be studied, first as to the general conditions of their having any meaning, which is the *Grammatica Speculativa* of Duns Scotus, secondly, as to the conditions of their truth, which is logic, and thirdly, as to the conditions of their transferring their meaning to other signs.[3] The Sign, in general, is the third member of a triad; first a thing as thing, second a thing as reacting with another thing; and third a thing as representing another to a third. Upon a

careful analysis, I found that all these triads embody the same three conceptions, which I call after Kant, my Categories. I first named them Quality,
Relation and Representation.[4] I cannot tell you with what earnest and long-
continued toil I have repeatedly endeavored to convince myself that my
notion that these three ideas are of fundamental importance in philosophy
was a mere deformity of my individual mind. It is impossible; the truth of
their principle has ever reappeared clearer and clearer. In using the word
relation I was not aware that there are relations which cannot be analyzed
into relations between pairs of objects. Had I been aware of it, I should have
preferred the word *Reaction.* It was also perhaps injudicious to stretch the
meaning of the word Representation so far beyond all recognition as I did.
However, the words Quality, Reaction, Representation, might well enough
serve to name the conceptions. The *names* are of little consequence; the
point is to apprehend the conceptions. And in order to avoid all false associations, I think it far the best plan to form entirely new scientific names for
them. I therefore prefer to designate them as *Firstness, Secondness,* and
Thirdness. I will endeavor to convey to you some idea of these conceptions.
They are ideas so excessively general, so much more general than ordinary
philosophical terms, that when you first come to them they must seem to you
vague.

Firstness may be defined as follows: It is the mode in which anything
would be for itself, irrespective of anything else, so that it would not make
any difference though nothing else existed, or ever had existed, or could
exist. Now this mode of being can only be apprehended as a mode of feeling.
For there is no other mode of being which we can conceive as having no
relation to the possibility of anything else. In the second place, the First must
be without parts. For a part of an object is something other than the object
itself. Remembering these points, you will perceive that any color, say
magenta, has and is a positive mode of feeling, irrespective of every other.
Because, firstness is all that it is, for itself, irrespective of anything else,
when viewed *from without* (and therefore no longer in the original fullness of
firstness) the firstnesses are all the different possible sense-qualities,
embracing endless varieties of which all we can feel are but minute fragments. Each of these is just as simple as any other. It is impossible for a
sense quality to be otherwise than absolutely simple. It is only complex to
the eye of comparison, not in itself.

A *Secondness* may be defined as a modification of the being of one subject, which modification is *ipso facto* a mode of being of quite a distinct subject, or, more accurately, secondness is that in each of two absolutely

severed and remote subjects which pairs it with the other, not for my mind nor for, or by, any mediating subject or circumstance whatsoever, but in those two subjects alone; so that it would be just the same if nothing else existed, or ever had existed, or could exist. You see that this Secondness in each subject must be secondary to the inward Firstness of that subject and does not supersede that firstness in the least. For were it to do so, the two subjects would, in so far, become one. Now it is precisely their twoness all the time that is most essential to their secondness. But though the secondness is secondary to the firstness, it constitutes no limitation upon the firstness. The two subjects are in no degree one; nor does the secondness belong to them taken together. There are two Secondnesses, one for each subject; but these are only aspects of one Pairedness which belongs to one subject in one way and to the other in another way. But this pairedness is nothing different from the secondness. It is not mediated or brought about; and consequently it is not of a comprehensible nature, but is absolutely blind. The aspect of it present to each subject has no possible *rationale.* In their *essence,* the two subjects are not paired; for in its essence anything is what it is, while its secondness is that of it which is another. The secondness, therefore, is an accidental circumstance. It is that a blind reaction takes place between the two subjects. It is that which we experience when our will meets with resistance, or when something obtrudes itself upon sense. Imagine a magenta color to feel itself and nothing else. Now while it slumbers in its magenta-ness let it suddenly be metamorphosed into pea green. Its experience at the moment of transformation will be secondness.

The idea of Thirdness is more readily understood. It is a modification of the being of one subject which is a mode of a second so far as it is a modification of a third. It might be called an inherent reason. That dormitive power of opium by virtue of which the patient sleeps is more than a mere word. It denotes, however indistinctly, some reason or regularity by virtue of which opium acts so. Every law, or general rule, expresses a thirdness; because it induces one fact to cause another. Now such a proposition as, Enoch is a man, expresses a firstness. There is no reason for it; such is Enoch's nature,—that is all. On the other hand the result that Enoch dies like other men, as result or effect, expresses a Secondness. The necessity of the conclusion is just the brute force of this Secondness. In Deduction, then, Firstness by the operation of Thirdness brings forth Secondness. Next consider an Induction. The people born in the last census-year may be considered as a sample of Americans. That *these* objects should be Americans has no reason except that that was the condition of my taking them into consideration.

There is Firstness. Now the Census tells me that about half those people were males. And that this was a necessary result is almost guaranteed by the number of persons included in the sample. There, then, I assume to be Secondness. Hence we infer the *reason* to be that there is some virtue, or occult regularity, operating to make one half of all American births male. There is Thirdness. Thus, Firstness and Secondness following have risen to Thirdness.

There are my three categories. I do not ask you to think highly of them. It would be marvellous if young students in philosophy should be able to distinguish these from a flotsam and jetsam of the sea of thought that is common enough. Besides, I do not ask to have them distinguished. All thought both correct and incorrect is so penetrated with this triad, that there is nothing novel about it, and no merit in having extracted it. I do not at present make any definite assertion about these conceptions. I only say, here are three ideas, lying upon the beach of the mysterious ocean. They are worth taking home, and polishing up, and seeing what they are good for.

I will only say this. There is a class of minds whom I know more intimately probably than many of you do, in whose thought, if it can be called thought, Firstness has a relative predominance. It is not that they are particularly given to hypothetic inference, though it is true that they are so given; but that all their conceptions are relatively detached and sensuous. Then there are the minds whom we commonly meet in the world, who cannot at all conceive that there is anything more to be desired than power. They care very little for inductions, as such. They are nominalists. They care for the things with which they react. They do reason, so far as they see any use for it; and they know it is useful to read. But when it comes to a passage in which the reasoning employs the letters A, B, C, they skip that. Now the letters A, B, C are pronouns indispensible to thinking about Thirdness; so that the mind who is repelled by that sort of thought, is simply a mind in which the element of Thirdness is feeble. Finally, there is the geometrical mind, who is quite willing that others should snatch the power and the glory so long as he can be obedient to that great world-vitality which is bringing out a cosmos of ideas, which is the end toward which all the forces and all the feelings in the world are tending. These are the minds to whom I offer my three Categories as containing something valuable for their purpose.

These Categories manifest themselves in every department of thought; but the advantage of studying them in formal logic is, that there we have a subject which is very simple and perfectly free from all doubt about its premises, and yet is not like pure mathematics confined entirely to purely hypothetical premises. It is the most abstract and simple of all the positive sciences, and the correct theory of it is quite indispensible to any true metaphysics.

... Among the classes which ordinary logic recognizes, *general* classes are particularly important. These general classes are composed not of real objects, but of possibilities, and hence it is that the nominalist for whom (though he be so mild a nominalist as Hegel was) a mere possibility which is not realized is nothing but what they call an "abstraction," and little better, if at all, than a fiction. It will be instructive, therefore, to inquire what it is in the logic of relatives which takes the logical position occupied in ordinary logic by "generality," or in medieval language by the *universal.*

Let us see, then, what it is that, in the logic of relatives, corresponds to generality in ordinary logic. From the point of view of Secondness, which is the pertinent point of view, the most radical difference between systems is in their multitudes. For systems of the same multitude can be transformed into one another by mere change of Thirdness, which is not true of systems of different multitude.[5] A system of multitude *zero* is no system, at all. That much must be granted to the nominalists. It is not even a quality, but only the abstract and germinal possibility which antecedes quality. It is at most being *per se.* A system of multitude unity is a mere First. A system of the multitude of two is like the system of truth and falsity in necessary logic. A system of the multitude of three is the lowest perfect system. The finite multitudes are all marked by this character that if there be a relation in which every individual in such a system stands to some other but in which no third stands to that other, then to every individual of the system some other individual stands in that relation. We next come to the multitude of all possible different finite multitudes, that is to the multitude of the whole numbers. A system of this multitude, which I call the *denumeral* multitude, is characterized by this, that though finite, yet its individuals have what I call *generative relations.* These are dyadic relations of relate to correlate such that, taking any one of them, whatever character belongs to the correlate of that relation whenever it belongs to the relate, and which also belongs to a certain individual of the system, which may be called the *origin* of the relation, belongs to every individual of the system. For example, in the system of cardinal numbers from *zero* up, the relation of being next lower in the order of magnitude is a generative relation. For every character which is such that if it belongs to the number next lower than any number also belongs to the number itself, and which also belongs to the number *zero* belongs to every number of the system. The next multitude is that of all possible collections of different finite collections.

This is the multitude of irrational quantities. I term it the *first abnumeral* multitude. The next multitude is that of all possible collections of collections of finite multitudes. I call it the *second abnumeral multitude.* The next is the multitude of all possible collections of collections of collections of finite multitudes. There will be a denumeral series of such abnumeral multitudes.[6] I prove that these are all different multitudes in the following way. In the first place, I say, that taking any such collection, which we may designate as the collection of A's, if each individual A has an identity distinct from all other As, then it is manifestly true that each collection of As has an identity distinct from all other collections of As; for it is rendered distinct by containing distinctly different individuals. But the individuals of a denumeral collection, such as all the whole numbers, have distinct identities. Hence, it follows that the same is true of all the abnumeral multitudes. But this does not prove that those multitudes are all different from one another. In order to prove that, I begin by defining, after Dr. Georg Cantor, what is meant by saying that one collection of distinct objects, say the Bs, is greater in multitude than another multitude of distinct objects, say the As. Namely, what is meant is that while it is possible that every A should have a distinct B assigned to it exclusively and not to any other A, yet it is not possible that every B should have a distinct A assigned to it exclusively and not to any other B.[7] Now then suppose the As to form a collection of any abnumeral multitude, then all possible collections of different As will form a collection of the next higher abnumeral multitude. It is evidently possible to assign to each A a distinct collection of As for we may assign to each the collection of all the other As. But I say that it is impossible to assign to every collection of As a distinct A. For let there be any distribution of collections of As which shall assign only one to each A, and I will designate a collection of As which will not have been assigned to any A whatever. For the As may be divided into two classes, the first containing every A which is assigned to a collection containing itself, the second containing every A to which is assigned a collection not containing itself. Now, I say that collection of As which is composed of all the As of the second class and none of the first has no A assigned to it. It has none of the first class assigned to it for each A of that class is assigned to but one collection which contains it while this collection does not contain it. It has none of the second class assigned to it for each *A* of this class is assigned only to one collection which does not contain it, while this collection does contain it. It is therefore absurd to suppose that any collection of distinct individuals, as all collections of abnumeral multitudes are, can have a multitude as great as that of the collection of possible collections of its individual members.

But now let us consider a collection containing an individual for every individual of a collection of collections comprising a collection of every abnumeral multitude. That is, this collection shall consist of all finite multitudes together with all possible collections of those multitudes, together with all possible collections of collections of those multitudes, together with all possible collections of collections of collections of those multitudes, and so on *ad infinitum*. This collection is evidently of a multitude as great as that of all possible collections of its members. But we have just seen that this cannot be true of any collection whose individuals are distinct from one another. We, therefore, find that we have now reached a multitude so vast that the individuals of such a collection melt into one another and lose their distinct identities. Such a collection is *continuous*.

Consider a line which returns into itself,—a ring

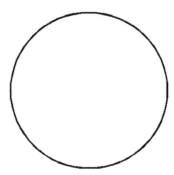

That line is a collection of points. For if a particle occupying at any one instant a single point, moves until it returns to its first position, it describes such a line, which consists only of the points that particle occupied during that time. But no point in this line has any distinct identity absolutely discriminated from every other. For let a point upon that line be marked

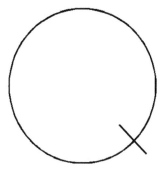

Now this mark is a discontinuity; and therefore I grant you, that this point is made by the marking distinctly different from all other points. Yet cut the line at that point

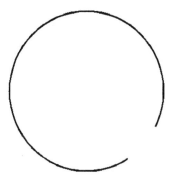

and where is that marked point now? It has become two points. And if those two end[s] were joined together so as to show the place,—they would become one single point. But if the junction ceased to have any distinguishing character, that is any discontinuity, there would not be any distinct point there. If *we* could not distinguish the junction it would not appear distinct. But the line is a mere conception. It is nothing but that which it can show; and therefore it follows that if there were no discontinuity there would *be* no distinct point there,—that is, no point absolutely distinct in its being from all others. Again going back to the line with two ends, let the last point of one end burst away

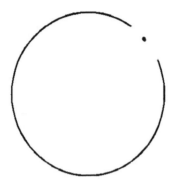

Still there is a point at the end still, and if the isolated point were put back, they would be one point. The end of a line might burst into any discrete multitude of points whatever, and they would all have been one point before the explosion. Points might fly off, in multitude and order like all the real irrational quantities from 0 to 1; and they might all have had that order of succession in the line and yet all have been at one point.[8] Men will say this is self-contradictory. It is not so. If it be so prove it. The apparatus of the logic of relatives is a perfect means of demonstrating anything to be self-contradictory that really is so; but that apparatus not only absolutely refuses to pronounce this self-contradictory but it demonstrates, on the contrary, that it is not so. Of course, I cannot carry you through that demonstration. But it is no matter of *opinion*. It is a matter of plain demonstration. Even although I should have fallen into some subtle fallacy about the series of abnumeral multitudes, which I must admit possible in the sense in which it is possible that a man might add up a column of five figures in all its 120 different orders and always get the same result, and yet that result might be wrong, yet I say, although all my conclusions about abnumerals were brought to ruin, what I now say about continuity would stand firm. Namely, a continuum is a collection of so vast a multitude that in the whole universe of possibility there is not room for them to retain their distinct identities; but they become welded into one another. Thus the continuum is all that is possible, in whatever dimension it be continuous. But the general or universal of ordinary logic also comprises whatever of a certain description is possible. And thus the *continuum* is that which the logic of relatives shows the *true* universal to be. I say the *true* universal; for no realist is so foolish as to maintain that *no* universal is a fiction.

Thus, the question of nominalism and realism has taken this shape: Are any continua real? Now Kant, like the faithful nominalist, that Dr. Abbot has shown him to be,[9] says no. The continuity of Time and Space are merely subjective. There is nothing of the sort in the real thing-in-itself. We are therefore not quite to the end of the controversy yet; though I think very near it.

What is reality? Perhaps there isn't any such thing at all. As I have repeatedly insisted, it is but a retroduction, a working hypothesis which we try, our one desperate forlorn hope of knowing anything.[10] Again it may be, and it would seem very bold to hope for anything better, that the hypothesis of reality though it answers pretty well, does not perfectly correspond to what is. But if there is any reality, then, so far as there is any reality, what that reality consists in is this: that there is in the being of things something which corresponds to the process of reasoning, that the world *lives,* and *moves,* and *has its being,* in [a] logic of events. We all think of nature as syllogizing. Even the mechanical philosopher, who is as nominalistic as a scientific man can be, does that. The immutable mechanical law together with the laws of attraction and repulsion form the major premise, the instantaneous relative positions and velocities of all the particles whether it be "at the end of the sixth day of creation,"—put back to an infinitely remote past if you like, though that does not lessen the miracle,—or whether it be at any other instant of time is the minor premise, the resulting accelerations form the conclusion. That is the very way the mechanical philosopher conceives the universe to operate.

I have not succeeded in persuading my contemporaries to believe that Nature also makes inductions and retroductions. They seem to think that her mind is in the infantile stage of the Aristotelian and Stoic philosophers. I point out that Evolution wherever it takes place is one vast succession of generalizations, by which matter is becoming subjected to ever higher and higher laws; and I point to the infinite variety of nature as testifying to her Originality or power of Retroduction.[11] But so far, the old ideas are too ingrained. Very few accept my message.

I will submit for your consideration the following metaphysical principle which is of the nature of a retroduction: Whatever unanalyzable element *sui generis seems* to be in nature, although it be not really where it seems to be, yet must really be [in] nature somewhere, since nothing else could have produced even the false appearance of such an element *sui generis.* For example, I may be in a dream at this moment, and while I think I am talking and you are trying to listen, I may all the time be snugly tucked up in bed and sound asleep. Yes, that may be; but still the very semblance of my feeling a reaction against my will and against my senses, suffices to prove that there really is, though not in this dream, yet somewhere, a reaction between the inward and outward worlds of my life.

In the same way, the very fact that there seems to be Thirdness in the world, even though it be not where it seems to be, proves that real Thirdness there must somewhere be. If the continuity of our inward and outward sense be not real, still it proves that continuity there really is, for how else should sense have the power of creating it?

Some people say that the sense of time is not in truth continuous, that we only imagine it to be so. If that be so, it strengthens my argument immensely. For how should the mind of every rustic and of every brute find it simpler to imagine time as continuous, in the very teeth of the appearances,—to connect it with by far the most difficult of all the conceptions which philosophers have ever thought out,—unless there were something in their real being which endowed such an idea with a simplicity which is certainly in the utmost contrast to its character in itself. But this something must be something in some sense *like* continuity. Now nothing can be like an element so peculiar except that very same element itself.

Of all the hypotheses which metaphysicians have ever broached, there is none which quarrels with the facts at every turn, so hopelessly, as does their favorite theory that continuity is a fiction. The only thing that makes them persist in it is their notion that continuity is self-contradictory, and *that* the logic of relatives when you study it in detail will explode forever. I *have* refuted it before you, in showing you how a multitude carried to its greatest possibility necessarily becomes continuous. Detailed study will furnish fuller and more satisfying refutations.

The extraordinary disposition of the human mind to think of everything under the difficult and almost incomprehensible form of a continuum can only be explained by supposing that each one of us is in his own real nature a continuum. I will not trouble you with any disquisition on the extreme form of realism which I myself entertain that every true universal, every continuum, is a living and conscious being, but I will content myself with saying that the only things valuable, even here in this life, are the continuities.

The *zero* collection is bare, abstract, germinal possibility. The continuum is concrete, developed possibility. The whole universe of true and real possibilities forms a continuum, upon which this Universe of Actual Existence is, by virtue of the essential Secondness of Existence, a discontinuous mark—like a line figure drawn on the area of the blackboard. There is room in the world of possibility for any multitude of such universes of Existence. Even in this transitory life, the only value of all the arbitrary arrangements which mark actuality, whether they were introduced once for all "at the end of the sixth day of creation" or whether as I believe, they spring out on every hand and all the time, as the act of creation goes on, their only value is to be shaped into a continuous delineation under the creative hand, and at any rate their only use for us is to hold us down to learning one lesson at a time, so that we may make the generalizations of intellect and the more important generalizations of sentiment which make the value of this world. Whether when we pass away, we shall be lost at once in the boundless universe of possibilities, or whether we shall only pass into a world of which this one is the superficies and which itself is discontinuity of higher dimensions, we must wait and see. Only if we make no rational working hypothesis about it we shall neglect a department of logical activity proper for both intellect and sentiment.

Endeavors to effectuate continuity have been the great task of the Nineteenth Century. To bind together ideas, to bind together facts, to bind together knowledge, to bind together sentiment, to bind together the purposes of men, to bind together industry, to bind together great works, to bind together power, to bind together nations into great natural, living, and enduring systems was the business that lay before our great grandfathers to commence and which we now see just about to pass into a second and more advanced stage of achievement. Such a work will not be aided by regarding continuity as an unreal figment, it cannot but be helped by regarding it as the really possible eternal order of things to which we are trying to make our arbitrariness conform.

As to detached ideas, they are of value only so far as directly or indirectly, they can be made conducive to the development of systems of ideas. There is no such thing as an absolutely detached idea. It would be no idea at all. For an idea is itself a continuous system. But of ideas those are most suggestive which detached though they seem are in fact fragments broken from great systems.

Generalization, the spilling out of continuous systems, in thought, in sentiment, in deed, is the true end of life. Every educated man who is thrown into business ought to pursue an avocation, a side-study, although it may be well to choose one not too remote from the subject of his work. It must be suited to his personal taste and liking, but whatever it is, it ought, unless his reasoning power is decidedly feeble, to involve some acquaintance with modern mathematics, at least with modern geometry, including topology, and the theory of functions. For in those studies there is such a wealth of forms of conception as he will seek elsewhere in vain. In addition to that, these studies will inculcate a strong dislike and contempt for all sham-reasoning, for all thinking made easy, for all attempts to reason without clothing conceptions in diagrammatic forms.

23

The Logic of Continuity

[Peirce 1898b] The last of Peirce's Cambridge Conferences Lectures complements the collection-theoretic definition of continuity given in selection 22 with a geometrical analysis based on Listing's contributions to topology. In proving his Census Theorem, Listing identifies a number of topological invariants, which Peirce uses in the latter part of this lecture to classify such important continua as space, time, and sensible quality. As these are applications, rather than explications, of continuity, they are omitted here.

Peirce's extensions and improvements of Listing's theorem had mostly to do with the consideration of various kinds of singularities or breaches of continuity.[1] Here topology impinges more directly on the very definition of continuity: points are subsumed, at the end of this excerpt, under the broader heading of singularities.

The first part of the lecture is a lead-in to Listing. Peirce's account of the branches of geometry, and of the results of Cayley and Klein, is similar to that in selection 16. He interrupts the topological train of thought to address two hard questions about his definition of continuity:

(a) Why can there be, for any infinite multitude M, a discrete collection (that is, a collection of distinct individuals) of multitude M, but no discrete collection obtained by aggregating discrete collections of all multitudes?

(b) How can it make sense to talk of a continuous collection as exceeding, in multitude, all discrete collections?

His answer to (a) is closely related to what is usually known as Cantor's Paradox.[2] The proposed aggregate would have a multitude, which as a consequence of the construction would be the greatest multitude; but by Cantor's Theorem there *is* no greatest multitude. His answer to (b) occasions a very important characterization of his continuum as a *potential aggregate*. This is obviously relevant to the conception of points in the earlier lecture; since a potential aggregate is in many ways more general than particular, this passage is also important for understanding Peirce's earlier remarks about "continuity in the logic of relatives." The answer to (b) is then that given any collection C of multitude M, a continuum *could* contain a collection of points whose multitude exceeded *M,* and in that extended sense the continuum has a greater multitude than *C.*

The relationship between continuous and discrete collections is analogous to that between the collection of all whole numbers and individual whole numbers. Each of the latter is determinate and can be completely counted; the former cannot be completed or completely counted—it is, like a continuous/potential aggregate, "indeterminate yet determinable." Yet this circumstance does not in any way prevent us from obtaining knowledge of the whole collection. One can question the analogy, since the whole numbers have a multitude (the lowest infinite one) and thus on Peirce's view *can* in some sense be completed and perhaps even counted (though not by us). Still the discussion is a brilliantly suggestive one, in which Peirce wrestles with deep issues in the neighborhood of the present-day distinction between sets and classes.[3]

Of all conceptions Continuity is by far the most difficult for Philosophy to handle. You naturally cannot do much with a conception until you can define it. Now every man at all competent to express an opinion must admit as it seems to me that no definition of continuity up to quite recent times was nearly right, and I maintain that the only thoroughly satisfactory definition is that which I have been gradually working out, and of which I presented a first *ébauche* when I had the honor of reading a paper here in Cambridge in 1892,[4] and of the final form of which I have given you sufficient hints in these lectures. But even supposing that my definition, which as yet has not received that sanction which can only come from the critical examination of the most powerful and exact intellects, is all wrong, still no man not in leading strings as to this matter can possibly think that there was anything like a satisfactory definition before the labors of Dr. Georg Cantor, which only began to attract the attention of the whole world about [1883].[5]

But after a satisfactory definition of continuity has been obtained the philosophical difficulties connected with this conception only begin to [be] felt in all their strength. Those difficulties are of two kinds. First there is the logical difficulty, how we are to establish a method of reasoning about continuity in philosophy? and second there is the metaphysical difficulty, what are we to say about the being, and the existence, and the genesis of continuity?

As to the proper method of reasoning about continuity, the dictate of good sense would seem to be that philosophy should in this matter follow the lead of geometry, the business of which it is to study continua.

But alas! the history of geometry forces upon us some sad lessons about the minds of men. That which had already been called the Elements of geometry long before the day of Euclid is a collection of convenient propositions concerning the relations between the lengths of lines, the areas of surfaces, the volumes of solids, and the measures of angles. It concerns itself only incidentally with the intrinsic properties of space, primarily only with the ideal properties of perfectly rigid bodies, of which we avail ourselves to construct a convenient system of measuring space. The measurement of a thing was clearly shown by Klein, twenty five years ago, to be always extrinsic to the nature of the thing itself. Elementary geometry is nothing but the introduction to *geometrical metric,* or the mathematical part of the physics of rigid bodies. The very early Greek geometers, I mean for example ,[6] who is said to have written the first *Elements,* I have no doubt, considered metric as the philosophical basis and foundation, not only of geometry, but of mathematics in general. For it is to be remarked that considerably the larger part of Euclid's Elements is occupied with algebra not with geometry; and since he and all the Greeks, had a much stronger impulse to get to the logical foundation of any object of study than we have, and since it is only the first book of Euclid in which the logic has been a matter of deep cogitation, it is plain that it was originally, at least, conceived that those geometrical truths in the first book of the Elements lay at the foundation even of algebra itself. But Euclid certainly, and in my opinion much earlier Greeks, had become acquainted with that branch of geometry which studies the conditions under which different rays of light indefinitely prolonged will intersect in common points or lie in common planes. There is no accepted name for this branch. It is sometimes called descriptive geometry; but that is in violent conflict with the principles of nomenclature, since descriptive geometry is the accepted name of a branch of geometry invented by Monge and so named by him,[7]—a branch closely allied to this other doctrine but not the same. Clifford called the branch of which we are speaking, Graphics[8] (which conveys no implication); other writers call it synthetic geometry (though it may be treated analytically), geometry of position (which is the name of something else), modern geometry (when in fact it is ancient), intersectional geometry (though projection plays as great a *role* as section in it), *projective* geometry (though section is as important as projection), perspective geometry, etc. I would propose the name *geometrical optic.* Euclid, I say, and earlier Greeks were acquainted [with] this geometrical optic. Now to any person of discernment in regard to intellectual qualities and who knows what the Greeks were, and especially what the Greek geometers were, and most particularly what Euclid was, it seems to me incredible that Euclid should have been acquainted with geometrical optic and not have perceived that it was more fundamental,—more intimately concerned with the intrinsic nature of space,—than metric is. And indeed *a posteriori* evidences that he actually did so are not wanting. Why, then, did Euclid not say a single word about this optic in his elements? Why did he altogether omit it even in cases where he must have seen that its propositions were indispensible conditions of the cogency of his demonstrations? Two possible explanations have occurred to me. It may be that he did not know how to prove the propositions of *optic* otherwise than by means of *metric;* and therefore, seeing that he could not make a thorough job, preferred rather ostentatiously and emphatically, (quite in his style in other matters) omitting all mention of optical propositions. Or it may be that, being a university professor, he did not wish to repel students by teaching propositions that had an appearance of being useless. Remember that even the stupendous Descartes abandoned the study of geometry. And why? Because he said it was *useless.* And this he said *a propos* of conic sections! That he should have thought conic sections useless, is comparatively pardonable. But that he the Moses of modern thinkers should have thought that a philosopher ought not to study useless things is it not a stain of dishonor on the human mind itself?

In modern times the Greek science of geometrical optic was utterly forgotten, all the books written about it were lost, and mathematicians became entirely ignorant that there was any such branch of geometry. There was a certain contemporary of Descartes, one Desargues, who rediscovered that optic and carried his researches into it very far indeed. He showed clearly and in detail the great utility of the doctrine in perspective drawing and in architecture, and the great economy that it would effect in the cutting of stones for building. On the theoretical side he pushed discovery to an advance of a good deal more than two centuries. He was a secular man. But he worked alone, with hardly the slightest recognition. Insignificant men treated him with vitriolic scorn. His works, though printed, were utterly lost and forgotten. The most voluminous historians of mathematics though compatriots did not know that such a man had ever lived, until one day Michel Chasles walking along the *Quai des Grands Augustins,* probably after a meeting of the *Institut,* came across and bought for a franc a MS copy of one of those printed books. He took it home and studied it. He learned from it the

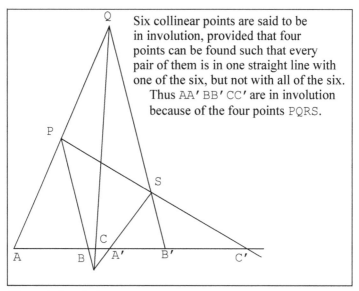

Six collinear points are said to be in involution, provided that four points can be found such that every pair of them is in one straight line with one of the six, but not with all of the six. Thus AA′ BB′ CC′ are in involution because of the four points PQRS.

Figure 1: Involution

important theory of the Involution of Six Points [(Fig. 1)]; and from him the mathematical world learned it; and it has been a great factor in the development of modern geometry. There can be no possible doubt that this knowledge actually came from the book of Desargues, because the relation has always borne the strange name Involution which Desargues had bestowed upon it,—and the whole theory is in his book although it had been totally unmentioned in any known treatise, memoir, or programme previous to the lucky find of Chasles. When will mankind learn the lessons such facts teach? That had that doctrine not been lost to all those generations of geometers, philosophy would have been further advanced today, and that the nations would have attained a higher intellectual level, is undoubtedly true,—but that may be passed by as a bagatelle. But why will men not reflect that but for the stupidity with which Desargues was met,—many a man might have eaten a better dinner and have had a better bottle of wine with it? It needs not much computation of causes and effects to see that that must be *so.*

In 1859, Arthur Cayley showed that the whole of geometrical metric is but a special problem in geometrical optic. Namely, Cayley showed that there is a locus in space,—not a *kind* of locus, but an individual place,— whose optical properties and relations to rigid bodies constitute those facts that are expressed by space-measurement.

That was in [.][9] It attracted the admiration and assent of the whole mathematical world, which has never since ceased to comment upon it, and develope the doctrine. Yet a few years ago I was talking with a man who had written two elementary geometries and who perhaps was, for aught I know still is, more influential than any other individual in determining how Geometry shall be taught in American schools at large, and this gentleman never heard of Projective Geometry neither the name nor the thing and his politeness never shone more than in his not treating what I said about Cayley with silent contempt.

But many years before Cayley made that discovery, a geometer in Göttingen, Listing by name,—a name which I will venture to say that Cayley, learned as he was in all departments of mathematics heard for the first time many years later, probably from Tait, who knew of him because he and Listing were both physicists,—this Listing had in 1847[,] four years before Riemann's first paper[,] discovered the existence of quite another branch of geometry, and had written two very long and rich memoirs about it.[10] But the mathematical world paid no heed to them till half a century had passed. This branch, which he called Topology, but which I shall call Topic, to rhyme with *metric* and *optic,* bears substantially the same relation to optic that optic bears to metric. Namely, *topic* shows that the entire collection of all possible rays, or unlimited straight lines, in space, has no general geometrical characters whatever that distinguish it at all from countless other families of lines. Its only distinction lies in its physical relations. Light moves along rays; so do particles unacted on by any forces; and maximum-minimum measurements are along rays. But the whole doctrine of geometrical optic is merely a special case of a topical doctrine.

That which *topic* treats of is the modes of connection of the parts of continua. *Geometrical topic* is what the philosopher must study who seeks to learn anything about continuity from geometry.

I will give you a slight sketch of the doctrine. We have seen in a previous lecture what continuity consists in. There is an endless series of abnumeral multitudes, each related to the next following as M is related to 2^M, where we might put any other quantity in place of 2. The least of these abnumeral multitudes is 2^N where N is the multitude of all whole numbers. It is impossible that there should be a collection of distinct individuals of greater multitude than all these abnumeral multitudes. Yet every one of these multitudes is possible and the existence of a collection of any one of these multitudes will not in the least militate against the existence of a collection of any other of these multitudes. Why then, may we not suppose a collection of distinct individuals which is an aggregate of one collection of each [of] those multitudes? The answer is, that to suppose an aggregate of *all* is to suppose the process of aggregation *completed,* and that is supposing the series of abnumeral multitudes brought to *an end,* while it can be proved that there is no last nor limit to the series. Let me remind you that by the *limit* of an endless series of successive objects we mean an object which comes after all the objects of that series, but so that every *other* object which comes after all those objects comes after the limit also. When I say that the series of abnumeral multitudes has no limit, I mean that it has no limit among multitudes of distinct individuals. It will have a limit if there is properly speaking any meaning in saying that something that is *not* a multitude of distinct individuals is *more* than every multitude of distinct individuals. But, you will ask, can there be any sense in that? I answer, yes, there can, in this way. That which is possible is in so far *general,* and as general, it ceases to be individual. Hence, remembering that the word "potential" means *indeterminate yet capable of determination in any special case,* there may be a *potential* aggregate of all the possibilities that are consistent with certain general conditions; and this may be such that given any collection of distinct individuals whatsoever, out of that potential aggregate there may be actualized a more multitudinous collection than the given collection. Thus the potential aggregate is with the strictest exactitude greater in multitude than any possible multitude of individuals. But being a potential aggregate only, it does not contain any individuals at all. It only contains general conditions which *permit* the determination of individuals.

The logic of this may be illustrated by considering an analogous case. You know very well that $2/3$ is not a whole number. It is not any whole number whatever. In the whole collection of whole numbers you will not find $2/3$. That you know. Therefore, you know something about the entire collection of whole numbers. But what is the nature of your conception of this collection? It is general. It is potential. It is vague, but yet with such a vagueness as permits of its accurate determination in regard to any particular object proposed for examination. Very well, that being granted, I proceed to the analogy with what we have been saying. Every whole number considered as a multitude is capable of being completely counted. Nor does its being aggregated with or added to any other whole number in the least degree interfere with the completion of the count. Yet the aggregate of *all* whole numbers cannot be completely counted. For the completion would suppose the *last* whole number was included, whereas there is no last whole number. But though the aggregate of all whole numbers cannot be completely counted, that does not prevent our having a distinct idea of the multitude of all whole numbers. We have a conception of the entire collection of whole numbers. It is a *potential* collection indeterminate yet determinable. And we see that the entire collection of whole numbers is more multitudinous than any whole number.

In like manner the potential aggregate of all the abnumeral multitudes is more multitudinous than any multitude. This potential aggregate cannot be a multitude of distinct individuals any more than the aggregate of all the whole numbers can be completely counted. But it is a distinct general conception for all that,—a conception of a potentiality.

A potential collection more multitudinous than any collection of distinct individuals can be cannot be entirely vague. For the potentiality supposes that the individuals are determinable in every multitude. That is, they are determinable as distinct. But there cannot be a distinctive quality for each individual; for these qualities would form a collection too multitudinous for them to remain distinct. It must therefore be by means of relations that the individuals are distinguishable from one another.

Suppose, in the first place, that there is but one such distinguishing relation, *r*. Then since one individual is to be distinguished from another simply by this that one is *r* of the other, it is plain that nothing is *r* to itself. Let us first try making this *r* a simple dyadic relation. If, then, of three individuals *A, B, C, A* is *r* to *B* and *B* is *r* to *C*, it must be that *A* is *r* to *C* or else that *C* is *r* to *A*. We do not see, at first, that there it matters which. Only there must be a general rule about it, because the whole idea of the system is the potential determination of individuals by means of entirely general characters. Suppose, first, that if *A* is *r* to *B* and *B* is *r* to *C* then in every case *C* is *r* to *A*, and consequently *A* is not *r* to *C*. Taken then any fourth individual, *D*. Either

A is *r* to *D* or *D* is *r* to *A*

If *A* is *r* to *D*, since *C* is *r* to *A*, *D* is *r* to *C*. Then either *B* is *r* to *D* or *D* is *r* to *B*		Either *C* is *r* to *D* or *D* is *r* to *C*	
		A is *r* to *C* *absurd*	Either *B* is *r* to *D* or *D* is *r* to *B*
C is *r* to *B* *absurd*	Since *B* is *r* to *C* *C* is *r* to *D* *absurd*		*C* is *r* to *B* *absurd*
			since *B* is *r* to *C* *C* is *r* to *D* *absurd*

That rule, then, when you come to look into it will not work. The other rule that if *A* is *r* to *B* and *B* is *r* to *C* then *A* is *r* to *C* leads to no contradiction, but it does lead to this, that there are two possible exceptional individuals one that is *r* to everything else and another to which everything else is *r*. This is like a limited line, where every point is *r* that is, is to the *right* of every other or else that other is to the right of it. The generality of the case is destroyed by those two points of discontinuity,—the extremities. Thus, we see that no perfect continuum can be defined by a dyadic relation. But if we take instead a triadic relation, and say *A* is *r* to B for *C*, say to fix our ideas that proceeding from *A* in a particular way, say to the right, you reach *B* before *C*, it is quite evident, that a continuum will result like a self-returning line

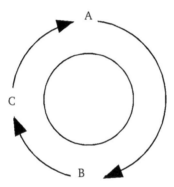

with no discontinuity whatever. All lines are simple rings and are topically precisely alike except that a line [may] have *topical singularities*. A *topical singularity* of a place is a place within that place from which the modes of departure are fewer or more than from the main collection of such places within the place. The topical singularities of *lines* are singular points. From an *ordinary* point on a line a particle can move two ways. Singular points are points from which a particle can move either no way, or in *one* way, or else in *three* ways or more. That is they are either, first, *isolated points* from which a particle cannot move in the line at all, or secondly, *extremities*, from which a particle can move but one way, or thirdly, *furcations*, from which a particle can move in three or more ways. Those are the only topical distinctions there are among lines. Surfaces, or two dimensional continua, can also have singularities. These are either *singular points* or *singular lines.* The singular lines are either isolated lines, which may have singular points at which they are not isolated, or they are *bounding* edges, or they are lines at which the surface splits into different sheets. These singular lines may themselves have singular points, which are subject [to] interesting laws. A student would find the singular lines of surfaces a good subject for a thesis. Isolated singular points of surfaces are either entirely detached from the surface or they are points at which different sheets or parts of the same sheet are tacked together. But aside from their singularities surfaces are of different kinds. In the first place, they are either *perissid* or *artiad.* A perissid surface is one which, although unbounded, does not enclose any space, that is, does not necessarily cut space into two regions, or what comes to the same thing, it has only one side. Such is the plane surface of geometrical optics, and in fact, such is every surface of odd order. The perissid surfaces are mathematically the simpler; but the artiad surfaces are the more familiar. A half twisted ribbon pasted together so that one side becomes continuous with the other side is an example of a bounded perissid surface. If you pass along a plane in geometrical optic, you finally come back to the same point, only you are on the other side of the plane.

An artiad surface, on the other hand, is for example the bounding surface between air and the stone of any finite stone, however curiously it may be cut. Moreover, a surface may have a *fornix* or any number of *fornices.* A fornix is a part of the surface like a railway-tunnel which at once bridges over the interval between two parts of the surface, and so connects them, and at the same time, tunnels under that bridge so that a particle may move on the surface from one side of the bridge to the other without touching the bridge. A flat-iron handle, or any handle with two attachments has a surface which is a *fornix* of the whole surface of which it forms a part. Both perissid and artiad surfaces can equally have any number of fornices, without disturbing their artiad or perissid character.

24

[On Multitudes]

[Peirce 1897(?)a] Peirce tries to come to grips, in this mathematically and philosophically abundant manuscript, with some of the difficult questions about continuity left open by the Cambridge Conferences Lectures. Here we find, in the opening summary of his theory, a full exposition of his subtly fallacious "proof" that a collection both supermultitudinous and discrete would violate Cantor's Theorem. Once the summary is done, Peirce takes on the first problem raised in selection 23, about the transition from discrete to continuous collections. In a series of ingenious constructions he tries to make out that as we move up the scale of his abnumeral multitudes even the elements of discrete collections begin to lose their identities.

The constructions are supposed to bring out the way in which infinitary interpolations of elements give rise to "incipient cohesiveness," even as the elements remain distinct. His main case study involves the placement of rational points on a line. Peirce tries to show that the placement of all those points (which are countably infinite), with the natural ordering, somehow carries with it the first abnumeral multitude of irrational points. The argument turns on the order of placement, and on the identification of rational points with individuals that can be specified by a finite number of characters. Peirce claims that if the interpolation process has a last step, and all of the rational points are present, then it is also somehow necessary that there be "along with them a first abnumeral multitude of individuals marked by denumerable series of characters." He never makes clear why, if a completed countable infinity of interpolations makes sense at all, it would not make sense to have a first interpolation after the countable one, yielding a countable infinity: why could we not mark, on a line, all of the rational points and then add one irrational point to obtain a counterexample to his claim about incipient cohesiveness?

Peirce suggests a more promising approach when he remarks that the rule that places the rationals in their natural order "necessarily assigns places not only to the denumerable collection but also to a collection of first abnumeral multitude of other possible individuals." Taken together as a whole, we might say, the rationals create the possibility of their limit points— including the irrational ones; and in general any distinct collection of points, of any multitude, creates the possibility of a larger multitude of interpolated points. This is one of the two key ideas of the most extended, and also the most successful and provocative, of the constructions in this selection. The other is that "number cannot possibly express continuity" but rather can only express discrete order. The construction begins with a countable infinity of rational approximations to π. On what has become the standard view—and on Peirce's own view in selection 19—the point representing π is determined as the limit of these approximations. Peirce's construction is a radical attack on this standard view, which he now rejects as question-begging. He maintains that so long as the collection of points on the line is discrete, it is possible to place an even larger collection on the line in between an isolated point and its neighbors. We do not run out of room so long as the points are distinct; that is, we do not fill up the line until the points have merged into one another, at which point we have a supermultitudinous collection of points.

The key ideas just identified are not invoked as premises in this construction; if anything, the construction supports rather than rests upon them. Ultimately Peirce's defense of the construction is that it is a "perfectly consistent hypothesis." But he also gives two geometrical arguments. The first of these points out that on the standard conception of the line the result of breaking it in two is to create two half-lines, exactly one of which has no endpoint; on Peirce's conception, where it makes sense to think of a single point splitting into two, both halves have an endpoint.

The second argument presupposes no more, Peirce claims, than "the intuitional idea of a line with which the synthetic geometer really works." Peirce imagines an infinite collection of blades cutting the line into pieces; so long as the blades are distinct from one another the result will be to chop the line not into points but rather into segments. In order to chop it into points, the blades must be in "incipient mergency." That is, they must be supermultitudinous. Moreover, since the number of points is no less than the number of segments into which the line can be divided, and any infinite multitude of blades chops the line into segments, the number of points on the line must be supermultitudinous.

Thus, the whole series of multitude, so far as yet made out begins with the multitude of a non-existent collection, or zero, and then comes the multitude of a single object, and then the multitude of 2, and so on increasing by one without end. After these multitudes comes the denumerable multitude which may be called the zero abnumeral multitude, then the first abnumeral multitude, then the second abnumeral, and so on increasing in order by one without end. All these multitudes thus form two denumerable series, and consequently there is only a denumerable multitude of different possible multitudes, so far as yet made out.

Let us now suppose that there is a collection of distinct objects of each of those multitudes. Then, taking any one of those collections, no matter what, there is, among the whole collection of those collections, a denumerable multitude of collections each of which is greater than the collection chosen. Let us then throw together all the distinct individuals of those collections so as to form an aggregate collection. This aggregate collection is greater than any of the single collections; for it has, as we have just seen, a denumerable collection of parts greater than any one of those collections. I shall call it a *supermultitudinous* collection.

It seems to be sufficiently evident that there is no collection at once greater than every abnumeral collection and less than such a supermultitudinous collection.

The collection of possible ways of distributing the individuals of a supermultitudinous collection, *S*, into two abodes is no greater than that supermultitudinous collection, *S*, itself. For denoting by *D* the denumerable multitude,[1] the abnumeral multitudes are

$$2^D \quad 2^{2^D} \quad 2^{2^{2^D}} \text{ etc.,}$$

or Exp *D*, $(\text{Exp})^2 D$, $(\text{Exp } D)^3 D$, etc.; and the magnitude of the supermultitudinous collection is the limit of this series. It is, in short, the result of a denumerable succession of exponential operations upon the denumerable multitude. But the magnitude of the collection of possible ways of distributing the individuals of a collection into two abodes is simply the result of an exponential operation upon the magnitude of the collection itself. Hence the magnitude of the ways of distributing the individuals of a supermultitudinous collection into two abodes is obtained by adding one more to the collection of exponential operations successively performed upon the denumerable multitude. But this collection of operations, being denumerable, the addition of one operation to it does not increase its magnitude. Hence, the collection of possible ways of distributing the individuals of a supermultitudinous collection into two abodes equals that collection itself.[2]

But we have already seen, Art. .,[3] that the multitude of ways of distributing the individuals of a collection of distinct individuals into two abodes exceeds the collection itself. Hence, it follows that a supermultitudinous collection is so great that its individuals are no longer distinct from one another.

This is not in conflict with the fact that a supermultitudinous collection is a denumerable collection of distinct collections in each of which the individuals are all distinct from one another.

A supermultitudinous collection, then, is no longer *discrete;* but is *continuous.* As such the term "multitude" ceases to be applicable to it; and I shall speak of its *order of magnitude,* meaning that character by which, considered as a collection, it is greater than one collection and less than another.

We have to consider whether or not all supermultitudinous collections are of the same order of magnitude. For that purpose, we need to develope a distinct notion of the relationship of such a collection to its individuals.

Let us begin with two individuals, which we may mark 0 and 1. We take a third individual; and beginning to construct a relation, r, which is to have the general property that if anything A is in the relation, r, to anything B, then B is not in the relation r to A (whence nothing is r to itself), we mark the third individual .1 and say that .1 is in the relation r to 0, while 1 is in the relation r to .1. We, next, add two more individuals, which we mark .01 and .11 [and] we say that .01 is r to 0, that .1 is r to .01, that .11 is r to .1 and that 1 is r to .11. We next add 4 new individuals, which we mark as .001, .011, .101 and .111. We say that .001 is r to 0 [and] that .01 is r to .001, that .011 is r to .01, that .10 is r to .011, that .101 is r to .10, that .11 is r to .101, that .111 is r to .11, and that 1 is r to .111. Our next addition will be of eight new individuals, .0001, .0011, .0101, .0111, .1001, .1011, .1101, .1111. Our next addition will be of sixteen new individuals. We go on until we have carried the additions as far as they can be carried without using an inenumerable number of characters to mark one individual.

Then all the individuals which are marked each by an enumerable collection of characters form a denumerable collection of individuals.

But now I call your attention to a very remarkable circumstance. All these individuals stand distinct and independent. Any one of them or any collection of them may be taken away without affecting the remainder; but yet there are already symptoms of incipient cohesiveness in them, a premonition of continuity. Remember that the multitude of grades of enumerable multitude is denumerable; and the multitude of all the grades of multitude is no greater. In dealing with finite numbers, where each multitude differs from the next but by one, we become accustomed to think that the multitude [of] *numerals,* or marks of grades of multitude, below a given numeral distinguishes that multitude from every other. But this ceases to be true when we leave the enumerable collections. If we use the term *arithm* to mean the multitude of grades of multitude below the multitude to which the arithm is attached, then the arithm of zero is zero, the arithm of 1 is 1, the arithm of 2 is 2, and in short the arithm of any enumerable multitude is that multitude. But the arithm of all higher multitudes is the same. It is the denumerable multitude, which may be called *infinity,* ∞. We shall, therefore, lose ourselves in a labyrinth [of] hopeless confusion if we allow ourselves for an instant to judge of a higher multitude by its arithm.

But I repeat that this unity of the arithms of all the higher multitudes is the first embryon of continuity. I proceed to explain this remark. In interpolating those individuals, I was just speaking of, the first interpolation being of 1 individual, the second interpolation being of 2 individuals, the third interpolation being of 4, the fourth interpolation being of 8 individuals, the fifth interpolation being of 16 individuals, and so on (I repeat this so many times in order to impress upon you what I here mean by the word interpolation), so long as there has been only an enumerable collection of interpolations, it is plain that not all of the individuals which are designated by enumerable collections of characters have as yet been inserted. The entire collection of those individuals that are marked by enumerable series of characters, remember, is *denumerable.* But as soon as the denumerable multitude of interpolations has been made, the collections of characters attached to the last inserted individuals are denumerable, and the collection of individuals there is of the first abnumeral multitude. Stop it at any point you please, however early, if all the individuals marked by enumerable series of characters are there, there is also along with them a first abnumeral multitude of individuals marked by denumerable series of characters. *They stick together.* This is what I meant by saying there was an incipient cohesiveness, a germin of continuity.

The cause of the phenomenon is so easily traced that hasty thinkers will say it is nothing but a fallacy. I do not agree with them. It is not true continuity, but only an appearance of cohesion; but in my opinion it is genuinely the first stage in the development of continuity.

The explanation which I allude to ought to be pretty obvious after what I have just said about the arithm. If the succession of insertions of individuals stops in such a way that there is a last inserted individual, this is either one of those individuals which are marked by enumerable collections of characters, in which case, not all of them have been inserted, or else it is one of those which are marked by denumerable collections of characters, in which case, in addition to the individuals marked by enumerable collections of characters, others amounting to the first abnumeral multitude have been inserted besides. But it is not necessary that the succession of insertions should be so broken off that there is any last individual inserted, and if this is the case, it may be that all the individuals marked by enumerable collections of characters have been inserted and no others. If all those individuals *are* inserted, the multitude of interpolations is precisely the same, namely denumerable, whether those interpolations do or do not include interpolations of the individuals marked by denumerable collections of characters.

Thus far, all the symptoms of cohesion which manifest themselves depend upon the order of succession. An enumerable collection does not cling together at all in whatever order it be taken. A denumerable succession does not cling together if taken in the order of its generation, any further than this, that no part can be struck off the latter end without so much is struck off as to reduce it to an enumerable collection. A denumerable collection can be so arranged that a denumerable collection of denumerable sequences can be struck out from it without reducing its multitude. Suppose for example we arrange the whole numbers in the following order.

1st, powers of the lowest prime in their order, 1, 2, 4, 8, 16, 32, 64, 128, 256, 512, 1024, etc.

2nd, each of these successively multiplied by powers of the next lowest prime in their order,

   3, 9, 27, 81, etc

   6, 18, 54, 162, etc

   12, 36, 128,[4] 324, etc

      etc

3rd, each of these successively multiplied by powers of the next lowest prime in their order

5, 25, 125, 625, etc	15, 75, 375, 1875, etc
10, 50, 250, 1250, etc	45, 225, 1125, 5625, etc
20, 100, 500, 2500, etc	135, 675, 3375, 16875, etc
etc	etc, etc.

and so on *ad infinitum*. . . .

There is also an abnumeral collection of arrangements of any denumerable collection such that it is impossible to reduce it to an enumerable collection without striking out more distinct parts than there are individuals remaining. Such an arrangement of the whole numbers is obtained for example by arranging them in what would be their natural order if the succession of figures with which they are written were reversed. . . .

For any such arrangement of a denumerable collection there must be some general rule connecting the place of an individual in the arrangement with its designation or symbol. *This rule necessarily assigns places not only to the denumerable collection but also to a collection of first abnumeral multitude of other possible individuals.* Thus a denumerable collection cannot be even in idea without the potential accompaniment of [a] first abnumeral collection.

And yet the constituent individuals of the abnumeral collection are distinct and discrete. Any one of them may be annihilated without affecting the others.

Observe that the precise manner in which the denumerable collection involves the first abnumeral collection is this, that in order to be able to say, here is complete the entire denumerable collection, it is necessary to have stopped somewhere and to have put down a last,—for by the word "complete," we mean to a last,—and before there can be any last after the denumerable collection is all there, a first abnumeral collection must already be there.

It is equally true that it is impossible to have complete the entire first abnumeral collection of sets of individuals of a denumerable collection without also having a second abnumeral collection; but this does not strike us so forcibly, because we never trouble ourselves to imagine that the entire first abnumeral collection of sets of denumerable individuals *is* complete. Another point of difference in the two cases is this, a denumerable collection seldom presents itself with a last. It *may* do so, as when the numbers are arranged

$$1, 3, 5, 7, 9, 11 \ldots 12, 10, 8, 6, 4, 2, 0$$

And there are arrangements in which there is no break at any particular point as

$$\frac{0}{1} \cdots \frac{1}{10} \cdots \frac{1}{9} \cdots \frac{1}{8} \cdots \frac{1}{7} \cdots \frac{1}{6} \cdots \frac{1}{5} \cdots \frac{2}{9} \cdots \frac{1}{4} \cdots \frac{2}{7} \cdots \frac{3}{10} \cdots \frac{1}{3} \cdots \frac{3}{8} \cdots \frac{2}{5} \cdots \frac{3}{7} \cdots \frac{4}{9} \cdots \frac{1}{2}$$

But in order that all the gaps may be *completely* filled it is necessary that infinitely high denominators should be used, and thus a first abnumeral collection is there. In order that the entire system of this first abnumeral collection should be there it is necessary that a second abnumeral collection should be there.

In order to show this, we must imagine the individuals of the first abnumeral collection so expressed as to give the imagination something to lay
hold upon in order to conceive of their being all there. Let us imagine for
example we have lists of all possible sets of rational quantities. The multitude of these lists is the first abnumeral. But in order to be able to say that the
collection of all possible such lists is *complete* they must be carried so far
that irrational quantities are included; and as soon as that is done, the multitude of lists is the second abnumeral.

We become habituated to think that numbers are capable in themselves
of expressing magnitude or at least proportional magnitude. If by magnitude
be meant *multitude,* of course this is so; but taking magnitude in the sense of
continuous magnitude, numbers in themselves can express neither magnitudes, nor the ratios of magnitudes. Numbers express nothing whatsoever
except order, *discrete order.* The fraction $7/12$ expresses nothing whatever
except something greater than $6/11$ and less than $8/13$, greater than $8/14$
and less than $6/12$ etc. In other words, the rules of arithmetic prescribe that
the values [of] fractions shall follow a certain sequence, but in regard to the
equality of the different parts into which the unit is cut, it can take no further
cognizance than to *reckon* them as all units on a par. The logic of number can
never be mastered until this idea is fully grasped.

Number cannot possibly express continuity. We can perfectly well mark
a point

———————————— • ————————————

to express π, or 3.14159 and drawing two horizontal lines to the right and
left of it say that the perimeters of inscribed polygons shall be measured off
on the left hand line and those of circumscribed polygons on the right hand
line. And thus we bring before our eyes what ought to be clear enough to the
eye of reason, that there is nothing in the nature of numbers to forbid the
interpolation of any multitude of quantities between all the approximations
of a convergent series and its limit. There is nothing about numbers which
can possibly forbid there being between the points representing all the values
of a convergent series and its limit, any abnumeral collection of other points
whatsoever.

On the contrary, just as much and just in the same way, as the supposition that the denumerable collection of rational points are *completely* present in a line involves the existence of a primipostnumeral[5] collection of irrational points, so the supposition that the system of irrational points on a line is *complete* involves the existence upon it of a secundipostnumeral collection of other points intermediate between the series and their limits.

No doubt that the ordinary conception of a *limit* namely that the limit of an increasing convergent series is the *least* quantity which is greater than all the approximations of the series, so far begs the question that it arbitrarily limits the system of possible quantities to a primipostnumeral multitude; and when it is then assumed that there is just one point for every assignable quantity, this completes the *petitio principii.*

I may here remark that the ordinary argument used by writers on the doctrine of limits about "assignable" quantities equally begs the question. For by an *assignable* quantity is meant a quantity to which numerical notation can indefinitely approximate. It is very easy by a careful analysis of the argument to convince oneself that nothing more is meant. Hence, that argument begins by assuming that the system of quantities is a primipostnumeral collection. But the only thing the argument is designed to do is to exclude a larger multitude of quantities. It is, therefore, completely illogical. However, there are forms of presentation of the doctrine of limits which are perfectly unobjectionable; and the doctrine has its value if rightly presented. I only say that the cruder forms of it, such as will be found in Newcomb's treatise,[6] are illogical and out of agreement with modern conceptions of quantity.

While I am on the subject of fallacies, I may as well notice a point which might possibly puzzle you. It is substantially proved by Euclid that there is but one assignable quantity which is the limit of a convergent series.[7] That is, if there is an increasing convergent series, A, and a decreasing convergent series, B, of which every approximation exceeds every approximation of A, and if there is no rational quantity which is at once greater than every approximation of A and less than every approximation of B, then there is but one surd quantity so intermediate. Now it might seem to you as if it followed that there was but one surd quantity intermediate between every pair of rational quantities so that the multitude of surds could not be greater than the multitude of rational quantities. But there is no end to a denumerable series and there are, therefore, no two adjacent rational quantities. There is one surd quantity and only one for each convergent series, calling two series the same if their approximations all agree after a sufficient number of terms, or if their difference approximates toward zero. But this is only to say

that the multitude of surds equals the multitude of denumerable sets of rational quantities, which is, as we have seen, the *primipostnumeral* multitude.

Going back to our representation of π

---

we remark that there is plenty of room to insert a *secundipostnumeral* multitude of quantities between the convergent series and its limit. Any one of
those quantities may likewise be separated from its neighbors, and we thus
see that between it and its nearest neighbors there is ample room for a *tertio-
postnumeral* multitude of other quantities, and so on through the whole
denumerable series of postnumeral quantities.

But if we suppose that *all* such orders of systems of quantities have been
inserted, there is no longer any room for so inserting any more. For to do so
we must select some quantity to be thus isolated in our representation. Now
whatever one we take, there will always be quantities of higher orders filling
up the spaces on the two sides.

We therefore see that such a supermultitudinous collection sticks
together by logical necessity. Its constituent individuals are no longer distinct and independent subjects. They have no existence,—no hypothetical
existence,—except in their relations to one another. They are not subjects,
but phrases expressive of the properties of the continuum.

From a line as it is usually conceived in analysis, that is, as a *primipostnumeral* succession of points, its extremity, which is its last point, may in
logical possibility be taken away; and when that is done the line is left without an extremity at that part. So whenever a line is severed in the middle one
of the parts will necessarily have an extreme point while the other will necessarily be left without any extremity.

But supposing a line to be a supermultitudinous collection of points,
nothing of the sort is logically possible. To sever a line in the middle is to
disrupt the logical identity of the point there, and make it two points. It is
impossible to sever a continuum by separating the connections of the points,
for the points only exist by virtue of those connections. The only way to
sever a continuum is to burst it, that is, to convert that which was one into
two.

It has hitherto been the opinion of mathematicians,—I speak only of those who are thoroughly acquainted with the most modern achievements [of] this particular branch of mathematical philosophy,—that the [collection of] points upon a line is of that multitude which I call *primipostnumeral*. But I hold this class of thinkers in such extraordinary esteem that I believe that when that opinion is refuted they will hold to it no longer. As for the swarm of pedagogues who infest this land, where pedagogy is so terribly overdone that instruction is generally supposed to be the chief purpose of a university, they have never heard of the opinion itself and never will hear of its refutation.

But that the collection of points upon a line is really supermultitudinous, is, I am confident, made evident by the following considerations. Across a line a collection of blades may come down simultaneously, and so long as the collection of blades is not so great that they merge into one another, owing to their supermultitude, they will cut the line up into as great a collection of pieces each of which will be a line,—just as completely a line as was the whole. This I say is the intuitional idea of a line with which the synthetic geometer really works,—his virtual hypothesis, whether he recognizes it or not; and I appeal to the scholars of this institution where geometry flourishes as all the world knows, to cast aside all analytical theories about lines, and looking at the matter from a synthetical point of view to make the mental experiment and say whether it is not true that the line refuses to be cut up into points by any discrete multitude of knives, however great. If this be the case the *lines* into which any line can be cut exceed any discrete multitude whatever. A line consists wholly of points, in one sense; for it is generated by a moving particle. But in order to chop a line up into its constituent points the blades of the chopper would have to be in incipient mergency into one another. They would have to be supermultitudinous; and so the points are supermultitudinous. Here then are two proofs. One is this:

The possible lines into which any line may be cut at one chop exceed any discrete multitude. Now the points on a line form a collection at least as great as the collection of the possible lines into which it can be chopped. Hence, the points of a line are supermultitudinous.

The other proof is this:

A line consists wholly of points; but in order to chop a line into points, the two ends of each piece must unite; and to do that without shrinkage they must merge into one another. Hence the collection of blades of the chopper must be so great that its constituent individual blades are no longer distinct.

In other words, they must be supermultitudinous; and the points into which this chopper severs the line must form an equal collection.

This I declare to be the synthetic geometer's hypothesis of the relation of a line to its points. But it does not affect my argument if it be not so. It is sufficient for my main purpose that it is a perfectly consistent hypothesis. For all I am trying to do is to elucidate the conception of a supermultitudinous collection and show that it involves no contradiction. In order to clinch my argument I am going presently to restate the matter in exact logical terms.

25

Infinitesimals

[Peirce 1900] Josiah Royce's *The World and the Individual* contained a long "Supplementary Essay" (Royce 1899, 473–588) on "The One, the Many and the Infinite," in which he correctly criticized Peirce for failing to recognize Cantor's opposition to infinitesimals. In this letter to the editor of *Science,* Peirce does not offer anything close to an adequate defense against this charge. He does, however, make some interesting remarks about Dedekind's definition of infinity, in comparison with his own; he also provides one of the best short summaries of the "supermultitudinous" conception of the continuum that he had been developing over the last four years or so.

This summary is very carefully laid out. Peirce first proves his version of Cantor's Theorem, and then "postulate[s] . . . [as] an admissible hypothesis" the existence of a line capable of supporting a point set of any multitude. (He complicates the statement of this last condition with additional stipulations on the betweenness relations on the line, designed to ensure that it has no endpoints, and returns upon itself.) He then demonstrates the incompatibility of these postulates with the Cantor/Dedekind view, on which the points of a continuous line can be put into one-to-one correspondence with the real numbers.

He goes on to argue that, according to his postulates, points cannot be regarded as constituent parts of the line. He gives his usual grounds for this: if points were constituents they would form a collection with a definite multitude, contradicting the postulate that the line can support an arbitrarily large point set. Though Peirce does not use the phrase 'potential aggregate,' his explanation of why all the points there could be on a line cannot form a collection hinges, like that in selection 23, on the merely potential nature of points. What is new here, relative to that earlier discussion, is a list of the ways in which individuals can be distinguished from one another, and a demonstration that merely potential points cannot be individuated in any of these ways.

These arguments sharpen the differences between continua and collections of distinct individuals, to the point where Peirce no longer speaks of the former as collections at all. To get to the bottom of those differences one would need to understand the ontological connection between collections and their members; so Peirce is in effect setting the agenda for the writings on collections excerpted here, all of which center around that very connection.[1] At the same time the denial that points are constituents of lines naturally prompts one to ask what it is for anything to be a constituent of something else; and the general problem of parts and wholes does in fact come to the fore in the next selection on continuity.

TO THE EDITOR OF *SCIENCE:* Will you kindly accord me space for a few remarks about Infinity and Continuity which I seem called upon to make by several notes to Professor Royce's Supplementary Essay in his strong work 'The World and the Individual'? I must confess that I am hardly prepared to discuss the subject as I ought to be, since I have never had an opportunity sufficiently to examine the two small books by Dedekind, nor two memoirs by Cantor, that have appeared since those contained in the second volume of the *Acta Mathematica.* I cannot even refer to Schröder's Logic.[2]

1. There has been some question whether Dedekind's definition of an infinite collection or that which results from negativing my definition of a finite collection is the best. It seems to me that two definitions of the same conception, not subject to any conditions, as a figure in space, for example, is subject to geometrical conditions, must be substantially the same. I pointed out (*Am. Journ. Math.* IV. 86, but whether I first made the suggestion or not I do not know) that a finite collection differs from an infinite collection in nothing else than that the syllogism of transposed [quantity] is applicable to it[3] (and by the consequences of this logical property). For that reason, the character of being finite seemed to me a positive extra determination which an infinite collection does not possess. Dr. Dedekind defines an infinite collection as one of which every *echter Theil* is similar to the whole collection.[4] It obviously would not do to say a *part,* simply, for every collection, even if it be infinite, is composed of individuals; and these individuals are parts of it, differing from the whole in being indivisible. Now I do not believe that it is possible to define an *echter Theil* without substantially coming to my definition. But, however that may be, Dedekind's definition is not of the kind of which I was in search. I sought to define a finite collection in logical terms. But a 'part,' in its mathematical, or collective, sense, is not a logical term, and itself requires definition.

2. Professor Royce remarks that my opinion that differentials may quite logically be considered as true infinitesimals, if we like, is shared by no mathematician '*outside of Italy.*'[5] As a logician, I am more comforted by corroboration in the clear mental atmosphere of Italy than I could be by any seconding from a tobacco-clouded and bemused land (if any such there be) where no philosophical eccentricity misses its champion, but where sane logic has not found favor. Meantime, I beg leave briefly to submit certain reasons for my opinion.

In the first place, I proved in January, 1897, in an article in the *Monist* (VII. 215), that the multitude of possible collections of members of any given collection whatever is greater than the multitude of the latter collection itself.[6] . . . That is, every multitude is less than a multitude; or, there is no maximum multitude.

In the second place I postulate that it is an admissible hypothesis that there may be a something, which we will call a *line,* having the following properties: 1st, points may be determined in a certain relation to it, which relation we will designate as that of 'lying on' that line; 2d, four different points being so determined, each of them is separated from one of the others by the remaining two; 3d, any three points, *A, B, C,* being taken on the line, any multitude whatever of points can be determined upon it so that every one of them is separated from *A* by *B* and *C.*

In the third place, the possible points so determinable on that line cannot be distinguished from one another by being put into one-to-one correspondence with any system of 'assignable quantities.' For such assignable quantities form a collection whose multitude is exceeded by that of another collection, namely, the collection of all possible collections of those 'assignable quantities.' But points are, by our postulate, determinable on the line in excess of that or of any other multitude. Now, those who say that two different points on a line must be at a finite distance from one another, virtually assert that the points are distinguishable by corresponding (in a one-to-one correspondence) to different individuals of a system of "assignable quantities." This system is a collection of individual quantities of very moderate multitude, being no more than the multitude of all possible collections of integral numbers. For by those 'assignable quantities' are meant those toward which the values of fractions can indefinitely approximate. According to my postulate, which involves no contradiction, a line may be so conceived that its points are not so distinguishable and consequently can be at infinitesimal distances.

Since, according to this conception, any multitude of points whatever are determinable on the line (not, of course, by us, but of their own nature), and since there is no maximum multitude, it follows that the points cannot be regarded as constituent parts of the line, existing on it by virtue of the line's existence. For if they were so, they would form a collection; and there would be a multitude greater than that of the points determinable on a line. We must, therefore, conceive that there are only so many points on the line as have been marked, or otherwise determined, upon it. Those do form a collection; but ever a greater collection remains determinable upon the line. *All* the determinable points cannot form a collection, since, by the postulate, if they did, the multitude of that collection would not be less than another multitude. The explanation of their not forming a collection is that all the determinable points are not individuals, distinct, each from all the rest. For individuals can only be distinct from one another in three ways: First, by acts of reaction, immediate or mediate, upon one another; second, by having *per se* different qualities; and third, by being in one-to-one correspondence to individuals that are distinct from one another in one of the first two ways. Now the points on a line not yet actually determined are mere potentialities, and, as such, cannot react upon one another actually; and, *per se,* they are all exactly alike; and they cannot be in one-to-one correspondence to any collection, since the multitude of that collection would require to be a maximum multitude. Consequently, all the possible points are not distinct from one another; although any possible multitude of points, once determined, become so distinct by the act of determination. It may be asked, "If the totality of the points determinable on a line does not constitute a collection, what shall we call it?" The answer is plain: the possibility of determining more than any given multitude of points, or, in other words, the fact that there is room for any multitude at every part of the line, makes it *continuous.* Every point actually marked upon it breaks its continuity, in one sense.

Not only is this view admissible without any violation of logic, but I find—though I cannot ask the space to explain this here—that it forms a basis for the differential calculus preferable, perhaps, at any rate, quite as clear, as the doctrine of limits. But this is not all. The subject of topical geometry has remained in a backward state because, as I apprehend, nobody has found a way of reasoning about it with demonstrative rigor. But the above conception of a line leads to a definition of continuity very similar to that of Kant. Although Kant confuses continuity with infinite divisibility, yet it is noticeable that he always defines a continuum as that of which every part (not every *echter Theil*) has itself parts. This is a very different thing from infinite divisibility, since it implies that the continuum is not composed of points, as, for example, the system of rational fractions, though infinitely divisible, is composed of the individual fractions. If we define a continuum as that every part of which can be divided into any multitude of parts whatsoever—or if we replace this by an equivalent definition in purely logical terms—we find it lends itself at once to mathematical demonstrations, and enables us to work with ease in topical geometry.

3. Professor Royce wants to know how I could, in a passage which he cites, attribute to Cantor the above opinion about infinitesimals.[7] My intention in that passage was simply to acknowledge myself, in a general way, to be no more than a follower of Cantor in regard to infinity, not to make him responsible for any particular opinion of my own. However, Cantor proposed, if I remember rightly, so far to modify the kinetical theory of gases as to make the multitude of ordinary atoms equal to that of the integral numbers, and that of the atoms of ether equal to the multitude of possible collections of such numbers.[8] Now, since it is essential to that theory that encounters shall take place, and that promiscuously, it would seem to follow that each atom has, in the random distribution, certain next neighbors, so that if there are an infinite multitude in a finite space, the infinitesimals must be actual real distances, and not the mere mathematical conceptions, like $\sqrt{-1}$, which is all that I contend for.

26

The Bed-Rock beneath Pragmaticism

[Peirce 1905a] In this footnote to an unfinished article for the *Monist,* Peirce promises, but does not deliver, an improved definition of continuity which begins with a clarification of the part/whole relation. According to his definition of 'material part' it makes sense to talk of the material parts of what we would call a time-slice of a jack-knife (a jack-knife at a specific time). A jack-knife, by contrast, can retain its identity even though it has time-slices whose material parts are not the same. Peirce drives this last point home with a reference to Jeannot's knife (the "jack-knife of the celebrated poser"), the proverbial French tale of the ship-of-Theseus-like jack-knife that never wore out because its handle and blade were constantly being replaced in alternation. Therefore, since material parts are constitutive of the whole whose parts they are, it makes sense to talk of a jack-knife's material parts only in a loose and extended sense. Peirce never gets around to applying this analysis to the definition of continuity—beyond stating at the outset that "[w]hatever is continuous has material parts"—though he does begin to define the "pseudo-continuum" of Cantor and others. It is not completely clear what his own definition would have looked like, but it would apparently still have involved multitude, which is the topic of a passage that degenerates into an embarrassingly bitter attack on the state of logic; as this passage adds nothing new (other than a direct reference to Bolzano), it is omitted here. It also appears, from the fragmentary text that Peirce managed to finish, that the theory of collections was supposed to play a role; for the discussion of material parts concludes with a definition of a collection as an object whose material parts "have no other connection between them than co-being." Peirce then begins his definition of a pseudo-continuum (that is, the Cantor/Dedekind continuum) by noting that it is "a collection of objects absolutely distinct from one another."

All of this sounds like the familiar prelude to the supermultitudinous view of continuity that has been the centerpiece of the last few selections. But there are discordant notes as well. Peirce rather plaintively remarks of the concept of collection that he has been "led . . . to believe it to be indecomposable," and even the definition of a pseudo-continuum gets (and stays) bogged down in complications over the distinctness of real numbers from one another. We will probably never know why Peirce never got any further than this. But it could well be because he was trying to pour new wine into old skins, with the usual results.

I feel that I ought to make amends for my blundering treatment of Continuity in a paper entitled 'The Law of Mind,' in Vol II of *The Monist,* by here redefining it after close and long study of the question. Whatever is continuous has *material parts.* I begin by defining these thus: The *material parts* of a thing or other object, W, that is composed of such parts, are whatever things are, firstly, each and every one of them, other than W; secondly, are all of some one internal nature (for example, are all places, or all times, or all spatial realities, or are all spiritual realities, or are all ideas, or are all characters, or are all relations, or are all external representations, etc.); thirdly, form together a collection of objects in which no one occurs twice over, and fourthly, are such that the Being of each of them, together with the modes of connexion between all sub-collections of them constitute the being of W. Almost everything which has material parts has different sets of such parts, often various *ad libitum.* Nothing which has an Essence (such as an essential purpose or use, like the jack-knife of the celebrated poser,) has any material parts in the strict sense just defined. But the term "material parts" may, without confusion (if a little care be exerted,) be used in a somewhat looser sense. Namely, if the Being (generally, a Concept) of an object, T, essentially involves something C which prevents it from having any material parts in the strict sense, and if there be something, W, which differs from T only in the absence of C and of any other such hindrances, so that W has material parts, then the material parts of W may loosely be termed material parts of T; but in such case, the concept of W so derived from T is nearly or quite always somewhat vague, so that either the material parts will be so, too, or else they must be conceived as merely the parts of some *state* of it, and very likely of an instantaneous state that is an *ens rationis* closely approximating to the nature of a fiction. It will be seen that the definition of Material Parts involves the concept of *Connexion,* even if there be no other connexion between them than co-being; and in case no other connexion be essential to the concept of W, this latter is called a *Collection,* concerning which I have merely to say that my reflexions on Mr. Alfred Bray Kempe's invaluable, very profound, and marvellously strong contribution to the science of Logic in the *Philosophical Transactions* for 1886, (which, by the way, seems to have proved too strong food for the mewling, etc. creatures who write the treatises on the science,) have led me to believe it to be indecomposable.[1] But I dare not be positive thereanent. . . .

But now I define a *pseudo-continuum* as that which modern writers on the theory of functions call a *continuum.* But this is fully represented by, and according to G. Cantor stands in one to one correspondence with the totality [of] real values, rational and irrational; and these are iconized, in their turn, according to those writers [by the] entire body of decimal expressions carried out to the right to all finite powers of $1/10$ without going on to Cantor's ωth place of decimals.

For it is a principle continually employed in the reasoning of the universally accepted "doctrine of limits" that two values that differ at all differ by a *finite value,* which would not be true if the ωth place of decimals were supposed to be included in their exact expressions; and indeed the whole purpose of the doctrine of limits is to avoid acknowledging that that place is concerned. Consequently the denumeral rows of figures which by virtue of a simple general principle are in one to one correspondence with the values, have relations among themselves quite regardless of their denoting those values that perfectly agree in form with the relations between the values; and consequently these unlimited decimal fractions themselves apart from their significations constitute a pseudo-continuum. This consideration renders it easy to define a pseudo-continuum. It is in the first place a collection of objects absolutely distinct from one another. Now from the fact that Cantor and others call it a "continuum," as well as from other things they say about it, I am led to suspect that they do not regard the pseudo-continuum of unlimited decimal expressions as all absolutely distinct from any other for the reason that taking any one of them it does not possess any one elementary and definite non-relative character which is not possessed by any other of them. But this is not what I mean, nor what is generally meant, by a collection of absolutely independent members. What I mean by that expression is that every member is distinguished from every other by possessing some one or another elementary and definite non-relative character which that other does not possess; and that this is the usual acceptation of the expression is evidenced by the fact that the majority of logicians are in the habit of conceiving of a universe of absolutely distinct individual objects by which they only mean that every individual is in every respect of a certain universe of respects determined in one or other of two ways and that every individual is differently determined from every other in *some* of those respects; and they do not generally conceive that every individual object has a determination in any one elementary and definite respect while all the other individuals are determined in the opposite way.

27

[Note and Addendum on Continuity]

[Peirce 1908e] In an important footnote to an installment in his "Amazing Mazes" series for the *Monist*—the first half of this selection—Peirce formulates a potentially fatal objection to his supermultitudinous theory of continuity. The objection has to do with linear orderings. Peirce shows how to impose such orderings, not just on countably infinite collections, but also on "first abnumerable" ones. But none of the strategies that are effective in these relatively simple cases are of any use with second abnumerable collections; nor is it clear that *any* strategies will work there, or with even larger collections. (The obvious connection with Zermelo's Axiom of Choice, and his Well-Ordering Theorem, is made explicit in selection 28.)[1] If no *n*th abnumerable multitude (*n* > 1) can be linearly ordered, then since the points on a line *are* linearly ordered, no collection of points larger than the real numbers can be placed on a line, and Peirce's supermultitudinous theory of the continuum collapses.

Peirce rightly points out that even if there is an upper bound on the multitude of points that can be placed on a line, it does not follow that a line can be filled with a point set of the appropriate multitude; and he appeals once again to our consciousness of time (in particular, to memory) to argue the need for a "more perfect continuity than the so-called 'continuity' of the theory of functions"; as in his supermultitudinous theory, "a line [with this more perfect continuity] does not consist of points."

By the time he received the proofs of the article, Peirce thought he could do better, and wrote three versions of an addendum for the published essay. The latest of the three, written on 26 May 1908, is included in this selection; it is the one that was completed and published. Peirce announces a new theory of continuity, based in topical geometry rather than the theory of collections. A true continuum obeys the (corrected) Kantian principle that every part has parts, and is such that all sufficiently small parts have the same mode of immediate connection to one another. Moreover, Peirce asserts, all the material parts (cf. selections 26 and 29) of a continuum have the same dimensionality. Rather than explaining the central idea of immediate connection, he notes that the explanation involves time, and answers the objection that his definition is therefore circular. It is perhaps an ominous sign that Peirce devotes to much space to what appears to be a somewhat manufactured objection: since he does not explain what he means by 'immediate connection,' it would hardly have occurred to the reader that time was bound up with such connection, had Peirce himself not brought it up. (In selection 29, the involvement of time in 'contiguity' is made clearer.) The excessive attention to side issues, when the main ideas are still so underexplained, would be less worrisome if Peirce had explained himself more fully elsewhere; but so far as we know, he did not.

Denumeral is applied to a collection in one-to-one correspondence to a collection in which every member is immediately followed by a single other member, and in which but a single member does not, immediately or mediately, follow any other. A collection is in one-to-one correspondence to another, if, and only if, there is a relation, r, such that every member of the first collection is r to some member of the second to which no other member of the first is r, while to every member of the second some member of the first is r, without being r to any other member of the second. The positive integers form the most obviously denumeral system. So does the system of all real integers, which, by the way, does not pass through infinity, since infinity itself is not part of the system. So does a Cantorian collection in which the endless series of all positive integers is immediately followed by ω_1, and this by ω_1+1, this by ω_1+2, and so on endlessly, this endless series being immediately followed by $2\omega_1$. Upon this follow an endless series of endless series, all positive integer coefficients of ω_1 being exhausted, whereupon immediately follows $\omega_1{}^2$, and in due course $x\omega_1{}^2 + y\omega_1 + z$, where x, y, z, are integers; and so on; in short, any system in which every member can be described so as to distinguish it from every other by a finite number of characters joined together in a finite number of ways, is a denumeral system. For writing the positive whole numbers in any way, most systematically thus:

1, 10, 11, 100, 101, 110, 111, 1000, 1001, 1010, 1011, etc.

it is plain that an infinite square matrix of pairs of such numbers can be arranged in one series, by proceeding along successive bevel lines thus:

(1,1);(1,10);(10,1) : (1,11);(10,10);(11,1) : (1,100);(10,11);(11,10) : etc.,

and consequently whatever can be arranged in such a square can be arranged in one row.

Thus an endless square of quaternions such as the following can be so arranged:

[(1,1)   (1,1)]:[(1,1)   (1,10)]; [(1,1)   (10,1)]:[(1,1)   (1,11)]; etc.
[(1,10) (1,1)]:[(1,10) (1,10)]; [(1,10) (10,1)]:[(1,10) (1,11)];etc.
[(10,1) (1,1)]:[(10,1) (1,10)]; [(10,1) (10,1)]:[(10,1) (1,11)]; etc.
[(1,11) (1,1)]:[(1,11) (1,10)]; [(1,11) (10,1)]:[(1,11) (1,11)]; etc.

Consequently whatever can be arranged in a block of any finite number of dimensions can be arranged in a linear succession. Thus it becomes evident that any collection of objects, every one of which can be distinguished from all others by a finite collection of marks joined in a finite number of ways can be of no greater than the denumeral multitude. (The bearing of this upon Cantor's ω^ω is not very clear to my mind.)[2] But when we come to the collection of all irrational fractions, to exactly distinguish each of which from all others would require an endless series of decimal places, we reach a greater multitude, or grade of maniness, namely, the *first abnumerable multitude.* It is called "abnumerable," to mean that there is, not only no way of counting the single members of such a collection so that, at last, every one will have been counted (in which case the multitude would be *enumerable*), but, further, there is no way of counting them so that every member will after a while get counted (which is the case with the single multitude called *denumeral*). It is called the *first* abnumerable multitude, because it is the smallest of an endless succession of abnumerable multitudes each smaller than the next. For whatever multitude of a collection of single members μ may denote, 2^μ, or the multitude of different collections, in such collection of multitude μ, is always greater than μ. The different members of an abnumerable collection are not capable of being distinguished, each one from all others, by any finite collection of marks or of finite sets of marks. But by the very definition of the first abnumerable multitude, as being the multitude of collections (or we might as well say of denumeral collections) that exist among the members of a denumeral collection, it follows that all the members of a first-abnumerable collection are capable of being ranged in a linear series, and of being so described that, of any two, we can tell which comes earlier in the series. For the two denumeral collections being each serially arranged, so that there is in each a first member and a singular next later member after each member, there will be a definite first member in respect to containing or not containing which the two collections differ, and we may adopt either the rule that the collection that contains, or the rule that the collection that does not contain, this member shall be earlier in the series of collections. Consequently a first abnumerable collection is capable of having all its members arranged in a linear series. But if we define a *pure* abnumerable collection as a collection of all collections of members of a denumeral collection each of which includes a denumeral collection of those members and excludes a denumeral collection of them, then there will be no two among all such pure abnumerable collections of which one follows next after the other or of which one next precedes the other, according to that rule. For example, among all decimal fractions whose decimal expressions contain each an infinite number of 1s and an infinite number of 0s, but no other figures, it is evident that there will be no two between which others of the same sort are not intermediate in value. What number for instance is next greater or next less than one which has a 1 in every place whose ordinal number is prime and a zero in every place whose ordinal number is composite? .11101010001010001010001000001 etc. Evidently, there is none; and this being the case, it is evident that all members of a pure second-abnumerable collection, which both contains and excludes among its members first-abnumerable collections formed of the members of a pure first-abnumerable collection, cannot, in any *such* way, be in any linear series. Should further investigation prove that a second-abnumeral multitude can in *no way* be linearly arranged, my former opinion that the common conception of a line implies that there is room upon it for any multitude of points whatsoever will need modification.

Certainly, I am obliged to confess that the ideas of common sense are not sufficiently distinct to render such an implication concerning the continuity of a line evident. But even should it be proved that no collection of higher multitude than the first abnumerable can be linearly arranged, this would be very far from establishing the idea of certain mathematico-logicians that a line consists of points. The question is not a physical one: it is simply whether there can be a consistent conception of a more perfect continuity than the so-called "continuity" of the theory of functions (and of the differential calculus) which makes the continuum a first-abnumerable system of points. It will still remain true, after the supposed demonstration, that no collection of points, each distinct from every other, can make up a line, no matter what relation may subsist between them; and therefore whatever multitude of points be placed upon a line, they leave room for the same multitude that there was room for on the line before placing any points upon it. This would generally be the case if there were room only for the denumeral multitude of points upon the line. As long as there is certainly room for the first denumerable multitude, no denumeral collection can be so placed as to

diminish the room, even if, as my opponents seem to think, the line is composed of actual determinate points. But in my view the unoccupied points of
a line are mere possibilities of points, and as such are not subject to the law
of contradiction, for what merely *can be* may also *not be.* And therefore
there is no cutting down of the possibility *merely* by some possibility having
been actualized. A man who can see does not become deprived of the power
merely by the fact that he has seen.

The argument which seems to me to prove, not only that there is such a
conception of continuity as I contend for, but that it is realized in the universe, is that if it were not so, nobody could have any memory. If time, as
many have thought, consists of discrete instants, all but the feeling of the
present instant would be utterly non-existent. But I have argued this elsewhere. The idea of some psychologists of meeting the difficulties by means
of the indefinite phenomenon of the span of consciousness betrays a complete misapprehension of the nature of those difficulties.

Added, 1908, May 26. In going over the proofs of this paper, written
nearly a year ago, I can announce that I have, in the interval, taken a considerable stride toward the solution of the question of continuity, having at
length clearly and minutely analyzed my own conception of a *perfect continuum* as well as that of an *imperfect continuum,* that is, a continuum having
topical singularities, or places of lower dimensionality where it is interrupted or divides. These labors are worth recording in a separate paper, if I
ever get leisure to write it. Meantime, I will jot down, as well as I briefly can,
one or two points. If in an otherwise unoccupied continuum a figure of lower
dimensionality be constructed,—such as an oval line on a spheroidal or
anchor-ring surface—either that figure is a part of the continuum or it is not.
If it is, it is a topical singularity, and according to my concept of continuity,
is a breach of continuity. If it is not, it constitutes no objection to my view
that all the parts of a perfect continuum have the same dimensionality as the
whole. (Strictly, all the *material,* or *actual,* parts, but I cannot now take the
space that minute accuracy would require, which would be many pages.)
That being the case, my notion of the essential character of a perfect continuum is the absolute generality with which two rules hold good, 1st, that
every part has parts; and 2d, that every sufficiently small part has the same
mode of immediate connection with others as every other has. This manifestly vague statement will more clearly convey my idea (though less distinctly,) than the elaborate full explication of it could. In endeavoring to
explicate "immediate connection," I seem driven to introduce the idea of
time. Now if my definition of continuity involves the notion of immediate

connection, and my definition of immediate connection involves the notion of time; and the notion of time involves that of continuity, I am falling into a *circulus in definiendo*. But on analyzing carefully the idea of Time, I find that to say it is continuous is just like saying that the atomic weight of oxygen is 16, meaning that that shall be the standard for all other atomic weights. The one asserts no more of Time than the other asserts concerning the atomic weight of oxygen;—that is, just nothing at all. If we are to suppose the idea of Time is wholly an affair of immediate consciousness, like the idea of royal purple, it cannot be analyzed and the whole inquiry comes to an end. If it can be analyzed, the way to go about the business is to trace out in imagination a course of observation and reflection that might cause the idea (or so much of it as is not mere feeling) to arise in a mind from which it was at first absent. It might arise in such a mind as a hypothesis to account for the seeming violations of the principle of contradiction in all alternating phenomena, the beats of the pulse, breathing, day and night. For though the *idea* would be absent from such a mind, that is not to suppose him blind to the *facts*. His hypothesis would be that we are, somehow, in a situation like that of sailing along a coast in the cabin of a steamboat in a dark night illumined by frequent flashes of lightning, and looking out of the windows. As long as we think the things we see are the same, they seem self-contradictory. But suppose them to be mere aspects, that is, relations to ourselves, and the phenomena are explained by supposing our standpoint to be different in the different flashes. Following out this idea, we soon see that it means nothing at all to say that time is unbroken. For if we all fall into a sleeping-beauty sleep, and *time itself stops during the interruption,* the instant of going to sleep is absolutely unseparated from the instant of waking; and the interruption is merely in our way of thinking, not in time itself. There are many other curious points in my new analysis. Thus, I show that my true continuum might have room only for a denumeral multitude of points, or it might have room for just any abnumeral multitude of which the units are in themselves capable of being put in a linear relationship, or there might be room for all multitudes, supposing no multitude is contrary to a linear arrangement.

28

Addition [on Continuity]

[Peirce 1908a] There are two manuscripts among Peirce's papers whose content links them closely to selection 27, and we know that the unpublished texts were written two days before the published one, because Peirce very considerately dated all three. He did not go so far as to note times of day, but the apparent progression of thought in the two unpublished manuscripts makes it likely that this one was written first. If so, then we can confidently say that on 24 May 1908 Peirce abandoned the supermultitudinous conception of continuity in favor of the one he articulates in selections 27 and 29. Selection 29 attempts a detailed exposition of the new conception announced publicly in selection 27. This manuscript, on the other hand, begins with what looks like a denial that a new conception is even called for. Recall that each of these three texts was meant to be appended to a note in which Peirce admits his uncertainty about the possibility of imposing a linear ordering on an arbitrarily large collection of points. In this fragmentary draft of that appendix he recants his uncertainty, only to have this renewed confidence evaporate—to judge from what he ultimately submitted to the printer—as he wrote. It is left to the reader to weigh the chronology, the content, and the incompleteness of these drafts against Peirce's claim, in the published addendum, that he has "at length clearly and minutely analyzed [his] own conception."

Aside from its bearing on that claim, the chief interest of this selection lies in what it tells us about Peirce's awareness of some of the more momentous mathematical developments of the day. In selection 27 his misgivings about linear orderings are tied to specific problematic collections, but here he voices more general worries about the connection between order and multitude. (It is possible that his doubts in selection 27 derive from that connection, but the text itself is silent on the point.) Peirce had a long-standing interest in the Trichotomy of Cardinals, which says that that given any two sets, either they have the same cardinality (multitude) or else one of them has a larger cardinality than the other (where cardinality is defined in terms of one-to-one correspondence). He made several attempts to prove this proposition, but by the time he wrote this manuscript he had come to recognize the difficulty and openness of the problem.[1] Trichotomy of Cardinals is equivalent to the Axiom of Choice, and therefore also to the Well-Ordering Theorem, which states that every set can be well-ordered, that is, admits of a linear ordering under which every non-empty subset has a least member. Ernst Zermelo (1871–1953) published his proof of the Well-Ordering Theorem (Zermelo 1904) four years before Peirce composed this text. Had Peirce known of that proof, and of the controversy it sparked, he would surely have mentioned it here. Instead he laments his inability to procure a copy of a proof of the Trichotomy of Cardinals, which he attributes to Émile Borel (1871–1956). Peirce is operating under a misunderstanding here. Borel's actual attitude towards Trichotomy was skeptical, and he was a leading opponent of the Axiom of Choice. He did, however, publish a proof, due to Felix Bernstein (1878–1956), of what is now known as the Schröder-Bernstein Theorem, in an appendix to (Borel 1898), so what Peirce says here may be a confused allusion to that.[2] It is unfortunate that Peirce does not elaborate on his reasons for thinking that Trichotomy of Cardinals implies the Well-Ordering Theorem. This fact was by no means universally recognized in 1908.[3] Since Peirce simply states the implication without explanation, there is no way of knowing, from this manuscript alone, whether the remark reflects his own insight into the problem, or whether he had gleaned it from somewhere else.

Addition, 1908 May 24. In reading the proofs of this article, which was written nearly a year ago, I find myself in a condition to take, as it seems to me, a long stride toward the solution of this important and dubious question of whether Cantor and Dedekind, followed by the general body of mathematicians[,] are right in holding the collective system of irrational and rational quantity to constitute a *continuum,* as I understand they do, or whether I have been right in maintaining that it should be called a *pseudo-continuum.* I still think, for the reasons given in *The Monist,* Vol. VII, pp. 205 *et seqq.,*[4] that there is room on a line for a collection of points of *any* multitude whatsoever, and not merely for a multitude equal to that of the different irrational values, which is, excepting one, the smallest of all infinite multitudes, while there is a denumeral multitude of distinctly greater multitudes, as is now, on all hands, admitted. I am obliged to grant, however, that the reasons to which I have just referred, being of the nature of logical analysis, and not of mathematical demonstration, leave us, in the present state of the science of Logic, not fully satisfied. I should, therefore, if I had be[en] able to do so, have

resorted to a proof by Borel, of the proposition that any two unequal collections stand in the same relation that any two unequal finite collections do,
since if they do, it seems to me clear that their units are inherently capable of
being put into a linear arrangement in every order of succession; and if this
be quite satisfactorily proved, I should be satisfied that there is room on a
line for a collection of points of any multitude. But to my vexation, I have
never been able to procure a copy of Borel's paper; and seeing that my reasons based on logical analysis seem to preclude the possibility of any mathematical demonstration, and knowing by my own experience the extreme
difficulty of either avoiding or detecting a vicious circle in attempting to
demonstrate that proposition, I still remain somewhat dubious about that.

But I wish to say now that while the view of Dedekind and Cantor
seems inconsistent with the hypothesis that after a point has been [inserted]
to denote each rational value between any two positive integers, or between
zero or infinity and one such value, or between zero and infinity, the order of
succession of the points on the line being the same as the order of values that
they severally represent (which would be easily enough done, if one could
accelerate his rate of working according to the proper law, and could mark a
mathematical point at all, and possessed the means of magnifying the line
indefinitely;) and if after that a point were inserted in the proper order of succession to denote each irrational value (but how this could be done, I cannot
in the least imagine,)

29

Supplement [on Continuity]

[Peirce 1908f] This is the second of Peirce's fragmentary attacks on the complex of problems he raises in selection 27. There is much more explicit attention to the part/whole relation in this version, including an elaborate taxonomy of the different kinds of parts, which Peirce never gets around to actually using in a definition of continuity. His general account of a part, as something which is necessarily present whenever its whole is, is perhaps more useful for the analysis of continuity than the detailed distinctions, which are based in Peirce's categories; the distinctions are hard to evaluate because the text breaks off before they are put to their intended use. (It is clear from what comes later that homogeneous parts would have played a central role, if Peirce had gotten that far.)

Peirce promises two definitions each of 'imperfect continuum' and 'perfect continuum' but only delivers one definition of the latter. Continuity is held to consist in a kind of regularity or homogeneity; this has a strongly logical flavor at first ("conformity to one Idea") but eventually the more "topical" idea of unbroken passage between contiguous parts becomes dominant. The role of time emerges more clearly here than in the later version: the "passage" from part to part is spelled out in temporal, indeed in quasi-mental, terms. Here the manuscript breaks off, perhaps because Peirce decided that it would be better to make the definition more thoroughly topical from the start. Or perhaps the published note gives us what would have been the continuation of this one; in that case this selection would be a first draft and not an alternative solution. As in selection 27, Peirce gets distracted just when it comes time to actually deliver his definition, in this instance by a charming but doubtfully relevant recollection from his boyhood.

This fragment and selection 27, taken together, give a somewhat mixed picture. There are enough common elements, and the elements cohere well enough, for there to be a discernible shift in Peirce's approach to the problem of continuity, and for some of the broad outlines of the new approach to be tolerably clear. At the same time, Peirce's tendency in these texts to break off altogeter at the moment of truth is more discouraging. It remains to be seen whether these intriguing but unfinished hints can be assembled and augmented so as to point clearly in a promising new direction.[1]

Supplement. 1908 May 24. In reading the proofs of this article, which was written nearly a year ago, I find myself in a condition to make, as it seems to me, a long stride toward clearing up the important question of whether Cantor and Dedekind, supported by an interpretation of Riemann's celebrated memoir on the hypotheses of geometry,[2] and followed, apparently, by the general body of mathematicians are right in holding that a collection of mathematical points each absolutely unextended and the collection being of the smallest but one of all infinite multitudes, and less than any of an endless series of multitudes each greater than those which precede it in the series, this collection being suitably arranged, constitutes a truly continuous line; or whether I am right in contending, *solus,* that though such a series of points no doubt has what is called continuity in the calculus and theory of functions, it has not the continuity of a line.

While Cantor's theory is that the continuity of space of all kinds, line[a]r, superfi[ci]al, and solid is constituted by there being this fewest of all multitude[s] of points, among those in which there are more units than in a simple endless series, so that it would be absolutely refuted the instant it was shown that there could be more numerous points upon a line, surface, or solid, my own theory nowise necessitates there being more, or upon there necessarily being any at all, although it does suppose that there is *room* for a good many (not necessarily, I think, even an infinite multitude). I proceed at once to define what I think it is that constitutes a true *continuum,* or continuous object.

I begin by defining a *part* of any *whole,* in a sense of the [term] much wider [than] any in current use, though it is not obsolete in the vocabulary of philosophy. In this broadest sense, [a part] is anything that is (1) other than its whole, and (2) is such that if the whole were really to be, no matter what else might be true, then the part must under all conceivable circumstance[s] itself really be, in the same 'universe of discourse,' though by no means necessarily in the same one of those three Universes with which experience makes us all more or less acquainted. Thus, light is a *part* of vision. The existence of the Campanile of San Marco is a *part* of the last sight I had of it; though the part was a thing while the whole is a mental experience. At any rate, so I, as an adherent of the doctrine of Immediate Perception, believe; and I might add that the same dear departed embodiment of beauty is a part of my present momentary recall of that experience. But it cannot be said that any man's progenitors are parts of himself; since it is conceivable that, instead of coming into the world by the mysterious process that he did, he should have been fabricated by the same fine art that Adam was, according to *Genesis* ii 7.

For the sake of defining different species of parts, which is requisite to making my definition of a continuum as distinct as I am, at present, able to make it, I must specify what my "three Universes of experience" are, and must, in such imperfect fashion as I can, indicate their principal departments. The first is the Universe of Ideas, i.e. of objects *in se* possible; the second is the Universe of Singulars, comprising physical Things and single Facts, or actualizations of ideas in singulars; while the third can, of course, only be the Universe of Minds with their Feelings, their Sensations of physical facts, and their Molitions (a word coined to express the putting forth of effort upon a physical thing, in contradistinction to the purpose of such action,) together with everything that pertains to Conduct (which is far too narrow a word,) Esthetic, Moral and Cognitive Impressions; Instincts; the formation of Plans, Purposes, and Ideals; Resolution the strengthening of Determinations, and Self-control; Habit-taking; the Formation, Utterance, and Interpretation of all sorts of mental Signs, whether Icons, Indices, or Symbols; mental Ejaculations, Commands and Interrogatories, and Judgments; Conjectures of Sympathy, Logical Analyses, and Testings; etc. By *real,* [I mean] possessing some characters independently of whether they have been attributed to that which is real by any individual human mind or any singular (i.e. neither indefinite nor yet general) group or other collection of minds, or not.[3] Thus, the substance of a dream or of a novel is not real; but the fact that the dream took place and that the novel was composed is real. By *existence* in any particular one of the three Universes of experience, I mean merely actual occurrence in the mode of Being common and peculiar to that universe. Yet I often use *exist* and *existence* (as I did above of the Campanile,) without qualification; and then I mean the actual Being of a physical thing. I apply to the word 'part' the four adjectives 'coexistential', '*copredicamental*', 'material', and 'homogeneous', to form a narrowing nest of classes of parts. All these adjectives except the one I have italicized are applied to 'part' in the books, but with such vague definitions as to disable them from service in the exacter philosophy of today. I venture to supply new definitions, that, I trust without serious violence to usage, may rescue them from the lumber-room. By a *coexistential* part, I mean a part that exists in the same one of "the three Universes of experience" as its whole. By a '*copredicamental*' part, I mean a part which belongs to the same '*predicament*', that is, in the same *summum genus* of the same Universe as its whole does, according to *true* enumeration of the predicaments. By a *non-partitional* character of a part, I mean one which is not conferred upon it by the partition. By a *material* part of a real whole, I mean a part which possesses all the real and non-partitional characters which are possessed in common at once by the whole in its unseparated state and by all the parts in an enumeration of all the mutually exclusive parts, which enumeration is such that no further partition into mutually exclusive parts could alter the real and non-partitional characters common to the whole and to all the parts. Very likely this definition might be greatly simplified; but I prefer its complexity to a risk of insufficiency. A *homogenous part* is a part which possesses among its real and non-partitional characters all those that belong to its whole and that a part can possess. A *homogeneous whole* is a whole entirely composed of homogeneous parts and possessing no other real character than such as is either possessed in common by all its homogeneous parts, or is one in respect to which those parts differ from one another, or is such as any whole of such homogeneous parts as its own must possess, or is the negation of a partitional character (that is, is the negation of a character not non-partitional, to avoid the undefined term 'partitional'), or is non-partitional.

I am now prepared to define that which I term a *perfect continuum* and that which I term an *imperfect continuum;* and I proceed to give two definitions of each, of which the first best expresses the essential character of its definitum but is not sufficiently explicit, while the second is couched entirely in carefully defined terms, but owing to the abstract character of the definitions being foreign to our usual way of thinking of continua, which we always perform by the aid of constructions in the imagination, is difficult to comprehend. The best way will be to study the second definition with the aid of examples, until one sees that it only amounts to a more explicit restatement of the first definition. The definitions of the imperfect continuum can give us no difficulty after those of the perfect continuum are once mastered.

One word as to the possibility of some error in the second definition. I remember, some fifty-odd years ago, a vessel struck an uncharted rock in Boston harbour, which had been an important port for more than two centuries during more than half of which it was the chief port of all the coast, lying in that part of the country where such matters as surveying received the most attention. I was only a boy; but I was with the eminent hydrographer who went to locate the rock; for bearings had been taken from the vessel when she struck, and considering the circumstances, they were good ones. Nevertheless, it was not until the middle of the third day's search that we found the rock. Now in endeavouring to make analyses like the drawing up of my second definition correct, I have often been struck with its being like the hunt for a sunken rock. The oversight if there be one, does not show itself on the surface, in the least, and one can only discover it by chancing, without any guide or clew, to hit upon the combination of circumstances under which the error shall make itself manifest. I have exercized my best care, and if there be any error, it may be merely that I have wrongly included under or excluded from the concept of continuity some kind of system so odd that nobody has yet thought of such a thing. If so, the matter can be set right later: the first question is whether I have struck the dominant note of the definition aright, or to drop the figure, whether my first definitions are substantially correct.

I now give my definitions, in the first place,

OF A PERFECT CONTINUUM

Definition I. A perfect continuum belongs to the genus, of a whole all whose parts without any exception whatsoever conform to one general law to which same law conform likewise all the parts of each single part. *Continuity* is thus a special kind of *generality*, or conformity to one Idea. More specifically, it is a *homogeneity*, or generality among all of a certain kind of parts of one whole. Still more specifically, the characters which are the same in all the parts are a certain kind of relationship of each part to all the coordinate parts; that is, it is a *regularity*. The step of specification which seems called for next, as appropriate to our purpose of defining, or logically analyzing the Idea of continuity, is that of asking ourselves what kind [of] relationship between parts it is that constitutes the regularity a continuity; and the first, and therefore doubtless the best answer for our purpose, not as the ultimate answer, but as the proximate one, is that it is the relation or relations of *contiguity;* for continuity is unbrokenness (whatever that may be,) and this seems to imply a *passage* from one part to a contiguous part. What is this "passage"? This passage seems to be an act of turning the attention from one part to another part; in short an actual event in the mind. This seems decidedly unfortunate, since an event can only take place in Time, and Time is a continuum; so that the prospect is that we shall rise from our analysis with a definition of continuity in general in terms of a special continuity. However, it is possible that this objection will disappear as we proceed.

Notes

1. The very most notable of these, in my view, is Peirce (1897), the second of the long reviews of Schröder (1880–1905) that Peirce published in the *Monist* at a very fertile time in his mathematical and philosophical development. I have also, less momentously, had to exclude from selection 12 an intriguing passage which ties Peirce's analysis of the natural numbers to the logic of the existential graphs. This is clearly of historical interest, for purposes of comparison between Peirce's mathematical writings on arithmetic and those of Dedekind and Frege; its philosophical ramifications are also worth exploring. But it would have necessitated a very thorough exposition of the existential graphs.

1. Paul Weiss gives a vivid description of Peirce's early education in a much-quoted passage: "From time to time they would play rapid games of double dummy together, from ten in the evening until sunrise, the father sharply criticizing every error. . . . The father's main efforts, however, were directed towards Charles's mathematical education. Rarely was any general principle or theorem disclosed to the son. Instead, the father would present him with problems, tables, or examples, and encouraged him to work out the principles for himself" (Weiss 1934, 399). Benjamin Peirce's profound and wide-ranging influence on his son is detailed in Peterson (1955).

2. Hookway's summary (Hookway 1985, 181) of the contents of *New Elements of Mathematics* gives a good sense of Peirce's mathematical range: "Alongside studies in mathematical logic and foundational issues, we find discussions of a wide range of topics: drafts of textbooks employing novel ideas of how the subject should be taught; mathematical studies of map projection deriving from his work for the United States Coastal Survey; discussions of linear algebra, probability, the four-color problem, the theory of measurement, non-Euclidean geometry." Fisch (1981, 383–385) gives a nice chronology of Peirce's major mathematical activities, concluding with what

he sees as the philosophical agenda that Peirce's mathematical writings set for us. Carolyn Eisele's essays, collected in Eisele (1979g), contain a wealth of detail on Peirce's mathematical interests and accomplishments; see also the General Introduction to Peirce (1976, xvi–xxvii) and Eisele (1995).

3. According to Fisch (1967, p. 193), Peirce's earliest use of this phrase dates to around 1893, in the *Grand Logic*. Fisch cites Ms 410, which contains a reference on page 17 to the views of the "extreme realist." Boler (1963) is a classic account of the content and origins of Peirce's later realism; Mayorga (2007) is a more recent treatment.

4. Boler (1963, 118) remarks that "when Peirce is through with the history of philosophy, there are very few realists left in it"; and Hookway writes that even Kant was eventually "demoted from being the champion of the realist conception of philosophy to standing as yet one more victim of nominalist illusion" (Hookway 1985, 288). It should already give us pause, as we seek to understand Peirce's realism, that he could ever have regarded *Kant* as "the champion of the realist conception of philosophy."

5. Peirce's views on the proper relationships between mathematics, logic and metaphysics are insightfully expounded in de Waal (2005).

6. I allude here to two of the three "global properties"—genericity and modality—of Peirce's continuum that are highlighted by Fernando Zalamea in his "succinct modern presentation" (Zalamea 2003, 139–154). The third is reflexivity, which we will come to in a moment.

7. These are two applications among many; Herron (1997, 594–597) lists them, along with Peirce's evolutionary cosomology, whose connection with continuity is treated in Alborn (1989) and Locke (2000). Other applications include: the proof of the pragmatic maxim (Zalamea forthcoming); laws of nature (Sfendoni-Mentzou 1997); and the existential graphs (Zeman 1968; Zalamea forthcoming).

8. Levy (1991) provides a good introduction to Peirce's thinking about infinitesimals, with useful discussions of the underlying motivations (127–132) and of Peirce's theory of higher-order infinitesimals (136–140). Havenel (2008, 101–103) and Herron (1997, 611–612) document the evolution of Peirce's attitude towards infinitesimals. Herron (1997) provides a comprehensive overview of his various approaches to the infinitely small.

9. My use of this quote might give a misleading impression of the passage whence it comes, which has been omitted from selection 20. It does not, however, mislead as to Peirce's view. So here, for the record, is the whole context: "Now if [as Peirce has just argued] exactitude, certitude, and univer-

sality are not to be attained by reasoning, there is certainly no other means by which they can be reached."

Straightforward though fallibilism seems at first blush, it turns out to be surprisingly tricky to state precisely: see Haack (1979, 46–54) for a review of the problems and a proposed solution. Both Haack and Elizabeth Cooke (2003) argue that despite his apparent ambivalence on the subject, there is no reason for Peirce not to extend his fallibilism to mathematics.

10. There is an obvious connection here to one of the leitmotivs of Peirce's later thought, the injunction that "deserves to be inscribed upon every wall of the city of philosophy," namely: "Do not block the way of inquiry." In the fourth Cambridge Conferences lecture, written just a few years after selection 20, Peirce lists four philosophical obstacles to the acceptance of this maxim. The first is infallibilism ("absolute assertion") and the fourth is "maintaining that this, that or the other element of science is . . . utterly inexplicable" (Peirce 1992, 178–180).

11. In his preface to the Hackett reissue of *The Development of Peirce's Philosophy,* Murphey takes a less pessimistic view of Peirce's later system as a whole (Murphey 1961, v). However this seems not to be because he takes a more optimistic view of Peirce's theory of the continuum: see Flower and Murphey (1977).

12. Susan Haack lists some of Peirce's interesting divergences from more familiar philosophies of mathematics in Haack (1993, 50–51).

13. This philosophical isolation is perhaps especially marked in connection with mathematics. Peirce does comment on some logicists, chiefly Dedekind and, to a lesser extent, Russell: see Hawkins (1997) on the latter, and Haack (1993, 33–35) for some of the most important references to the logicists in Peirce's writings. Peirce's fragmentary remarks on Russell's Paradox have been seen as a misfire (Murphey 1961, 241–242) and also as genuinely insightful (Hawkins 1997, 138–139).

Aside from these brushes with logicism, Peirce shows little if any awareness of what Stewart Shapiro (2000) calls "the big three": logicism, intuitionism and formalism. For helpful comparisons of Peirce's views with those of the better-known schools of thought, see Pietarinen (forthcoming) and Tiercelin (forthcoming). Both authors are concerned, as I am here, to identify ways in which Peirce can help to resolve stubborn tensions in the more standard approaches: Pietarinen argues that Peirce's pragmaticism is a distinctive brand of anti-foundationalism, which reconciles formalism and intuitionism; and Tiercelin highlights, in a treatment to which my own is much indebted, Peirce's realism as an alternative to Quine that is less vulnerable to Maddy's objections. Tiercelin's earlier paper on Peirce's mathematical realism (Tiercelin 1993) brings out, as does her later one as well, the unique mixture of platonism and conceptualism in his thinking about mathematics.

14. In the syllabus for the Lowell Lectures Peirce calls *The Phenomenology of Mind* "a work far too inaccurate to be recommended to any but mature scholars, though perhaps the most profound ever written" (Peirce 1903h, E2.266).

15. This is verbatim from Russell (1957, 71) except that the original has 'whether' in place of 'how we know that'. The change is necessary for my purposes because most philosophers (Russell included) take it for granted that we do know a good deal of mathematics: the problem is not whether, but how, we know it. In context the quote is much less pessimistic than it sounds when taken in isolation; Russell is in fact stating a version of Peirce's view that mathematics is the study of hypothetical states of things.

16. On the role of causal theories in Benacerraf's paper, see Burgess and Rosen (1997, 35–41). W. D. Hart recasts it as a problem for empiricism in the first four pages of his introduction to Hart (1996). For discussions of Benacerraf's Dilemma and Peirce's philosophy of mathematics, see Kerr-Lawson (1997, 79–81), and also the essays by Cooke, Hookway and Tiercelin in Moore (forthcoming 1).

17. One of Peirce's principal arguments for the objectivity of mathematics is its capacity to surprise us: "mathematical phenomena are just as unexpected as any results of chemical experiments. . . . Although mathematics deals with ideas and not with the world of sensible experience, its discoveries are not arbitrary dreams but something to which our minds are forced and which were unforeseen" (S6, 41). Susanna Marietti (forthcoming) argues that Peirce's semiotically grounded analysis of mathematical reasoning enables him to reconcile this feature of mathematics (which she dubs its *fertility*) with the certainty and universality of its results.

18. The technical mistakes are the assumption, as an obvious fact about infinite cardinals, of the Generalized Continuum Hypothesis; and a fallacious proof (S24, 191) that every infinite cardinal can be obtained, by a finite number of exponentiations, from the cardinal of the natural numbers. Herron (1997, 625–627) sketches a set theory in which the latter result holds good, and points out its unattractive features. See Murphey (1961, 286–288) for a comparison of Brouwer's intuitionism with Peirce's ontologically freewheeling approach to set-theoretic infinities. Putnam (1992, 46) compares Peirce's "iterative hierarchy" of collections with the now standard hierarchy

of ZFC (Zermelo-Fraenkel set theory with Choice), and also compares Peirce with Brouwer (50).

19. The Peirce/logicism (dis)connection has been something of a lightning rod in the secondary literature. The exchange between Haack (1993) and Houser (1993), over whether Peirce can be said to be, in any sense, a logicist, is a good place to start. Grattan-Guinness (1997) is especially helpful (33–34) on the significance, for Peirce's philosophy of mathematics, of his roots in the algebraic tradition in logic. Dipert (1995, 46) traces Peirce's rejection of logicism to his "view that all thought and communication is 'diagrammatic'."

20. De Waal (2005, 290–293) summarizes Peirce's views on the proper relationship between mathematics and logic; see also Grattan-Guinness (1997). Levy (1997, 88–95) argues that Peirce was not altogether consistent on the point, allowing that mathematics was after all dependent upon logic in many respects, and insisting on its independence mainly out of filial piety (109). Van Evra (1997) is a focused study of an important stage in the development of Peirce's views on the question. Shannon Dea (2006) finds a particularly radical affirmation of the priority of mathematics to logic in one of Peirce's book reviews.

21. The argument given here only supports the claim that science commits us to mathematical objects, but the consensus that mathematical objects are abstract is all but universal: "Mathematics tells [us] of infinities upon infinities of mathematical objects and of perfectly straight lines and extensionless points. Neither mind nor matter embodies things as numerous or as perfect" (Resnik 1997, 82).

22. See Roberts (1973) and Shin (2002).

23. The proof in Euclid provides partial confirmation of this claim: it proceeds by extending the sides of the triangle through the vertices of the base angles, and constructing new points on the resulting segments. Heath, in commenting on the family of proofs to which Peirce's belongs, remarks that "assuming the construction . . . of another triangle equal in all respects to the given triangle . . . is not in accordance with Euclid's principles and practice" (Heath 1956, 254).

The seminal paper on the theorematic/corollarial distinction is Hintikka (1983), which reconstructs the distinction in terms of modern quantification theory, and identifies it with Hintikka's own distinction between surface and depth tautologies. Hintikka's essay has spawned a fairly extensive literature. Other discussions include Hookway (1985, 193–201), Ketner (1985), Levy (1997), Shin (1997, 22–30), Marietti (2006, 122–124) and the concluding paragraphs of Shin (forthcoming); this is not an exhaustive list. For detailed examples of theorematic and corollarial reasoning, see Ketner (1985, 409–413), Levy (1997, 96–101), Shin (1997, 24–26), Campos (forthcoming), and Marietti (forthcoming). Hintikka (115) relates the distinction to that between analytic and synthetic judgments; Levy (1997, 104–106) goes into this at some length.

24. Maddy suggests that it is inappropriate to ask whether certain set theoretic axioms are true (Maddy 1997, 131–132). To rule out that question altogether is, with respect to those axioms at any rate, to give up on realism. As I interpret Peirce, he would not go so far, though he would insist that the question about truth must be properly understood.

25. Kerr-Lawson (1997, 82) compares Peirce's "bicategorial view" with Quine's "monotone ontology" and suggests that Peirce's approach to mathematical ontology is closer to that of most working mathematician.

26. See especially selections 7, 10, and 22.

27. Peirce claimed that the sufficiency of his three categories was established by a theorem showing that every relation of arity higher than three could be reduced to a compound of relations each having an arity of three or less. See Hookway (1985, 97–101) and also Short (2007, 74) for discussion and references to the literature.

28. The reality of Firstness also involves, for Peirce, the reality of objective chance, his commitment to which he calls his "tychism." Though tychism is an essential component of Peirce's wider architectonic, it does not play a major role in his philosophy of mathematics, so I will be giving it short shrift here.

29. In particular I have, among other things, illustrated the very general notion of an interpretant with the very special case of an interpreting thought. Peirce himself engaged in a similar simplification by calling the interpretant "an effect upon a person," a "sop to Cerberus, because I despair of making my broader conception understood" (Peirce 1908b, E2:478). For more on his broader conception, see Short (2007), especially chapter 7.

30. Boler (1963, 128) calls this "the most significant statement" of Thirdness for the purposes of understanding the admixture of idealism in Peirce's realism. As Boler observes, "the prototype of thirdness is found in the action of a sign"; so it is only to be expected that a very similar characterization of Thirdness is to be found in the correspondence with Lady Welby, which is primarily concerned with semiotics: "the mode of being of that which is such as it is, in bringing a second and a third into relation to each other" (Peirce 1904b, C8.328). Peirce's idealism and realism come together

in his affirmations that laws are "of the nature of a representation" (*HPL*, E2.181) and at the same time "really operative in nature" (E2.183). The passage just quoted bears close study, in conjunction with Boler's admirable exploration (131–138) of Peirce's objective idealism.

31. Hookway's fine discussion (Hookway 1985, 188–192) of selection 10 does not mention structuralism by name but clearly highlights the structuralist themes therein. He makes it explicit in Hookway (forthcoming).

32. Ahti-Veikko Pietarinen (forthcoming) has urged caution in identifying Peirce's structuralism with any of the varieties now current. In correspondence he has suggested some affinities between Peirce's views and the strands of contemporary structuralism that are heavily influenced by the mathematical theory of categories.

33. The exact phrasing is Putnam's (1979, 70). Putnam does not give a reference, but see Kreisel (1958, 138n1; 1965, 99). Tiercelin (forthcoming) is an important treatment of the objects/objectivity theme in Peirce's philosophy of mathematics.

34. There has been some controversy in the literature over whether Peirce meant this as praise or blame. I am persuaded by Mayorga's arguments on the side of praise (Mayorga 2007, pp. 77–78, 86–89).

35. The initial formulation of the definition (S12, 97) says that the being of the collection depends upon the *existence* of whatever has the defining quality; but Peirce goes on to allow that a collection can have essence, though lacking existence, if no such things exist.

36. Putnam, himself a leading exponent of modal-structuralism, ascribes some version thereof to Peirce in Putnam (1992, 50). Noble (1989) argues that Peirce's transition to what Potter calls the "Kantistic" conception of continuity resulted precisely from a prior change in his understanding of modality; Noble is especially illuminating on the interconnections, in Peirce's thinking, among modality, collections and continuity.

37. Peirce's philosophical analysis of the natural numbers in selection 15 is a major piece of evidence in favor of a structuralist interpretation of Peirce's mathematical ontology; I am much indebted to Hookway's discussion of that text, and of the role of abstraction therein, in Hookway (forthcoming).

38. Thomas Short (1988, 55n4) argues that the distinction is essential to a proper understanding of abstraction in empirical science, and that Peirce confuses the issue in selection 9. J. Jay Zeman (1983, 299), on the other hand, identifies the two notions. Peirce's usage is inconsistent, so the problem is not to be solved by a simple resort to the texts.

For an introduction to hypostatic abstraction in Peirce, see the works of Zeman and Short just cited, and also the tenth chapter of Short (2007). Hookway (1985, 201–203) is an excellent overview of abstraction as it relates to mathematics. Shin (forthcoming) applies Peirce's theory of abstraction to classical problems about generalization in the context of mathematical proof. Hintikka (1983, 114) notes in passing that Peirce himself "considered a vindication of the concept of abstraction as the most important application" of his corollarial/theorematic distinction. For more on the connection of abstraction with that distinction, see Zeman (295–297), Levy (1997, 101–102), and the conclusion of Shin (forthcoming).

39. Zeman (1983, 297–299) argues that for Peirce generally what counts as an abstraction is highly context-dependent. There is an obvious connection between this relativism and Peirce's account of perceptual judgments in *HPL* (E2.207–211), according to which "Thirdness pours in upon us through every avenue of sense" (211). It is surely no accident that this train of thought is set in motion by a discussion of the general elements that a mathematician must be able to *see* in her diagrams. Tiercelin (forthcoming) contains a very useful exploration of Peirce's "abductive logic of perception" and its implications for his mathematical epistemology. See also Hull (1994, 279–288), which makes explicit reference to the *HPL* discussion of perception and Thirdness.

40. The place of abstractions, and mathematical objects, in Peirce's categorial scheme is taken up by Murphey (1961, 239–240) and by Zeman (1983, 301–304).

41. "But most strictly speaking, even an idea, except in the sense of a possibility, or Firstness, cannot be an Icon. A possibility alone is an Icon purely by virtue of its quality; and its object can only be a Firstness. But a sign may be *iconic*, that is, may represent its object mainly by its similarity, no matter what its mode of being. If a substantive be wanted, an iconic Representamen may be termed a *hypoicon*. Any material image, as a painting, is largely conventional in its mode of representation; but in itself, without legend or label, it may be called a *hypoicon*. Hypoicons may roughly [be] divided according to the mode of Firstness which they partake. Those which . . . represent the relations, mainly dyadic, or so regarded, of the parts of one thing by analogous relations in their own parts, are *diagrams*" (Peirce 1903h, E2.273–274). Short (2007, 215–218) is a compact introduction to this portion of Peirce's semiotic taxonomy.

42. Campos (forthcoming) is an in-depth treatment of the role of semiotic imagination in mathematical reasoning, extending the analysis of exper-

imentation on iconic diagrams in Campos (2007, 474–477). Marietti (forthcoming) is another important discussion of the semiotics of diagrams, and of semiotic observation in mathematical reasoning; there and in Marietti (2005) she places great emphasis on the individuality, and the spatiotemporal character, of diagrams. There is also much valuable discussion of diagrammatic reasoning, and its semiotics, in Tiercelin (forthcoming). Though I have been giving pride of place to icons, as Peirce himself does, it is worth noting that symbols and indices also play an important part. Elizabeth Cooke (forthcoming) argues that indices in mathematical diagrams are a partial substitute for the Secondness that mathematical objects so conspicuously lack; Marietti (forthcoming) is especially enlightening on the interaction of symbols and icons in mathematical reasoning, a point she also expands on in Marietti (2006, 114–120). It is worth noting Dipert's observation (2004, p. 312) that we do not "yet understand clearly what exactly is iconic in the existential graphs and exactly what advantage this iconicity confers on diagrams."

Diagrams and their key role, not just in Peirce's philosophy of mathematics but also in his philosophy more generally, have rightly become a major focus in the literature. Stjernfelt (2000) is a very thorough treatment of diagrams and Peirce's theory of knowledge, with a great deal of semiotic analysis. Rosenthal (1984) explores the connection between Peirce's diagrammatic analysis of mathematical reasoning and his theory of meaning. J.E. Tiles (1988) argues, with Peirce and against Duhem and other deductivists, for the importance of iconic models in physical and well as mathematical science. Beverly Kent (1997) argues that diagrammatic thinking is a unifying source of Peirce's pragmatism, his existential graphs, and his classification of the sciences. Kathleen Hull (2005) makes a connection between diagrammatic reasoning and Peirce's "neglected argument" (Peirce 1908d) for theism.

The many discussions of diagrammatic reasoning in mathematics include, in addition to those already mentioned, Hookway (1985, 185–192) and Hoffmann (2003, 128–140). Campos (2009) explores the epistemic capacities that the mathematician must bring, on Peirce's view, to the task of diagrammatic reasoning; of particular note is the explicit emphasis, in Campos's closing case study of the exterior angle theorem (143–151), on the employment of these faculties by a *community* of mathematical inquirers. An important theme that has emerged from this burgeoning sub-literature is the link between the diagrammatic nature of mathematical reasoning and the ways in which mathematics, for Peirce, is a creative as well as scientific endeavor. Daniel Campos has pursued this in great depth in the papers just cited. Hull (1994) is another key source; and see Stroh (1985).

43. "The ground of interpretation of a mathematical theory when it is applied is iconic: the theory itself is a relational structure the elements of which are indices; and the theory is applicable to states of affairs containing elements involved in a relational structure of the same form. There is an iso-morphism between the mathematical theory and the reality to which it is applied" (Hookway 1985, 191). See Brady (2000, 113–115) for further dis-cussion of Peirce's view of mathematical reasoning as diagrammatic experi-mentation, and its relevance to ongoing philosophical debates over the scientific applicability of mathematics.

44. See, e.g., Parsons (1990, 289–292) and the works cited there.

45. On the implications of undecidable statements for Peirce's philoso-phy of mathematics, see Murphey (1961, 236–237).

1. [THE NATURE OF MATHEMATICS]

1. For further discussion of the Peirces' definitions of mathematics, see Murphey (1961, 229–237), Hookway (1985, 203–204), and Lenhard (2004, 94-100). Grattan-Guinness (1997, 34–35) argues that the upshot of Benjamin Peirce's definition is that necessary conclusions are those that are based on form alone. Paul Shields (1997, 43–44) suggests that Peirce's axiomatization of the natural numbers in Peirce (1881) was motivated in part by a desire to vindicate his father's definition. Campos (2007, 471–476) is a very useful analysis of "hypothetical states of things"; the essay as a whole explores the implications of Peirce's inclusion of hypothesis-making (*pace* his father) in the mathematician's brief.

2. Both of these examples are drawn from the theory of functions of a complex variable. The development of the complex plane (the "two-way spread of imaginary quantity") is recounted in Kline (1972, 628–632). Georg Friedrich Bernhard Riemann (1828–1866) invented the surfaces that bear his name as a geometrical representation of multiple-valued functions: see Kline (1972, 656–662) and Stewart and Tall (1983, 268–282).

3. As Peirce notes in selection 3 (16), Aristotle does not give a formal definition of mathematics. The discussions of mathematics Peirce probably has in mind here are in *Physics* II.2 (193b22–194b15) and *Metaphysics* XIII.3 (1077b18–1078b6). Aristotle distinguishes metaphysics (theology) as

first philosophy, prior to the other theoretical sciences (including mathematics), in *Metaphysics* VI, 1.

4. For more detail on Peirce's classification of the branches of geometry, see selections 16 (121–123) and 23 (181–184), and also Murphey (1961, 212–218).

5. In selection 4 Peirce will attribute this view to Boethius (475?–526?) and Ammonius (435?–517?). In *De Institutione Arithmeticae* (Masi 1983, 72–73) Boethius makes essentially the argument Peirce outlines here.

6. Augustus De Morgan (1806–1871) was a pioneer in the algebraic tradition in logic and a major influence on Peirce's logic; William Rowan Hamilton (1805–1865) was one of the most eminent mathematicians of the day and the discoverer of quaternions. For De Morgan's views, see De Morgan (1858, 76; 1860c, 184n1); for Hamilton's, see Hamilton (1837, 295, 297). Hamilton acknowledges the Kantian inspiration for his view in a letter to De Morgan written in 1841 and reprinted in Graves (1882–1889, vol. 3, 245–247). Pycior (1995, pp. 141–144) argues that Peirce's reading of De Morgan as an adherent of Hamilton's Kantianism is mistaken.

7. *Posterior Analytics* I.10 (76b31–34).

8. Peirce defines 'artiad', 'perissid' and 'singularity' in selection 23 (187f). Note that while 'artiad' and 'perissid' are defined there as properties of surfaces, Peirce asks in the text whether *space* is perissid. He extends the definitions from surfaces to spaces in (Peirce 1895(?)a, N3.123): "A space which contains no boundless surfaces but artiad surfaces is called an *artiad* space. A space which independently of singularities, etc., contains a boundless perissid surface is called a *perissid* space." For more on these concepts, see Eisele (1979b, 278–280). Havenel (forthcoming) contains an invaluable glossary of Peirce's topological vocabulary.

William Kingdon Clifford (1845–1879) made important contributions to non-euclidean geometry, and is well-known to philosophers as the author of "The Ethics of Belief"; he was one of the prominent scientists Peirce met during a European tour in 1875. Clifford raises the possibility that space contains singularities in Peirce's sense in Clifford (1886, 229).

9. *Critique of Pure Reason,* A80/B106 and A162/B202ff.

10. Peirce is quoting the first sentence of the first chapter of Peirce (1870, 2–4). The passage continues:

> This definition of mathematics is wider than that which is ordinarily given, and by which its range is limited to quantitative research. The ordinary definition like that of other sciences, is objective; whereas this is subjective.

Recent investigations, of which quaternions is the most noteworthy instance, make it manifest that the old definition is too restricted. The sphere of mathematics is here extended, in accordance with the derivation of its name to all demonstrative research, so as to include all knowledge strictly capable of dogmatic teaching. Mathematics is not the discoverer of laws, for it is not induction; neither is it the framer of theories, for it is not hypothesis; but it is the judge over both, and it is the arbiter to which either must refer its claims; and neither law can rule nor theory explain without the sanction of mathematics. It deduces from a law all its consequences, and develops them into the suitable form for comparison with observation and thereby measures the strength of the argument from observation in favor of a proposed law, or of a proposed form of application of a law. Mathematics, under this definition, belongs to every enquiry, moral as well as physical. Even the rules of logic, by which it is rigidly bound could not be deduced without its aid. The laws of argument admit of simple statement, but they must be curiously transposed, before they can be applied to the living speech and verified by observation. In its pure and simple form the syllogism cannot be directly compared with all experience, or it would not have required an Aristotle to discover it. It must be transmuted into all the possible shapes, in which reasoning loves to clothe itself. The transmutation is the mathematical process in the establishment of the law. Of some sciences, it is so large a portion that they have been quite abandoned to the mathematician which may not have been altogether to the advantage of philosophy. Such is the case with geometry and analytic mechanics. But in many other sciences, as in all those of mental philosophy and most of the branches of natural history, the deductions are so immediate and of such simple construction, that it is of no practical value to separate the mathematical portion and subject it to isolated discussion.

11. In Peirce (1903i, 260) the "three branches" of logic are enumerated as follows:

(1) Speculative Grammar, or the general theory of the nature and meanings of signs, whether they be icons, indices or symbols; (2) *Critic*, which classifies arguments and determines the validity and degree of force of each kind; (3) *Methodeutic*, which studies the methods that ought to be pursued in the investigation, in the exposition, and in the application of truth.

Peirce often names Duns Scotus as the originator of speculative grammar, but the treatise by that title included in Wadding's edition of Scotus is now known to be spurious (Boler 2004, 64).

12. Chrystal (1851–1911) was Professor of Mathematics at the University of Edinburgh. His entry (Chrystal 1883) begins: "Any conception which is definitely and completely determined by means of a finite number of spec-

ifications, say by assigning a finite number of elements, is a mathematical conception. Mathematics has for its function to develop the consequences involved in the definition of a group of mathematical conceptions." Peirce spells Chrystal's surname correctly in selection 3 (18).

13. Probably an allusion to Peirce's bitterly disappointing efforts to find a publisher for his innovative revision (excerpted here in selection 6) of his father's geometry text. For a detailed account see pp. xiv–xxvi of Eisele's introduction to Peirce (1976, Vol. 2).

14. Comte's original classification is in the second chapter of the *Cours de philosophie positive* (Comte 1830–1842, 35–67). The fourth chapter (80–139) of Kent (1987) surveys the many stages through which Peirce's classification passed, finally stabilizing, according to Kent, in 1903. For more on Peirce's mature architectonic, see Short (2007, 61–66).

2. The Regenerated Logic

1. The assumption Peirce objects to is—the probably spurious (Mueller 1981, 35)—Common Notion 5; it is invoked in the proof of Proposition I.16: "In any triangle, if one of the sides be produced, the exterior angle is greater than either of the interior and opposite angles" (Euclid 1956, 279–280). Their violation of this "axiom" was long viewed as a paradoxical feature of actual (as opposed to potential) infinities (Moore 1990, pp. 7, 54, 79). This is presumably why in rejecting it Peirce mentions Georg Cantor (1845–1918), the founder of transfinite set theory, and Augustin Louis Cauchy (1789–1857), whose approach to the foundations of analysis (Kline 1972, 948–965) Peirce saw as an affirmation of the reality of infinitesimals (S19, 152). Peirce is more explicit about the connection between actual infinities and the failure of the exterior angle theorem in selection 8 (62).

2. On the first of these controversies, see Laudan (1968); on the second, see Love (1927, 13–15).

3. In the entry for 'science' (Whitney 1889, 5397), written by Peirce.

3. The Logic of Mathematics In Relation to Education

1. Hoffmann (2003, 131–133) discusses the Kantian roots of Peirce's analysis of diagrams and diagrammatic reasoning. See Marietti (2005, 203–205) for more on this, and on Peirce's departures from Kant's approach.

2. Girard Desargues (1591–1661) and Michel Chasles (1793–1880) are major figures in the history of projective geometry. Peirce recounts the story in more detail in selection 23 (182); cf. Kline (1972, 288–289).

3. *Critique of Pure Reason* A714/B742f.

4. Ibid., A142/B182f.

5. Jevons (1864). Jevons's full alternative title is "The Logic of Quality apart from Quantity." Boole's "fundamental equation" is $x^2 = x$ (Boole 1854, 31) where x is one of Boole's "literal symbols ... representing things as subjects of our conceptions" (27). The equation says that the class of things falling under the concept (represented by) x is identical to the class of things falling under (x and x). Boole subsequently (49–51) derives the Law of Noncontradiction from this equation.

6. The "Theorem of Pappus" is more usually known as Desargues's Theorem. Following O'Hara and Ward (1937, 30–31) we say that the triangles ABC and $A'B'C'$ are *centrally perspective* just in case there is a point lying on all three of the lines AA', BB', and CC'; they are *axially perspective* just in case there is a line on which the intersection point of the lines AB and $A'B'$ lies together with the points of intersection of the other two pairs of lines determined by corresponding pairs of vertices. Desargues's Theorem says that if two triangles are centrally perspective, then they are axially perspective. Peirce discusses the theorem in the third of the Harvard Lectures on Pragmatism (Peirce 1903a, E2.173–174). Karl Georg Christian von Staudt (1798–1867) gives his proof in Staudt (1847, 41). Barycentric coordinates, a method of coordinatizing a projective plane, were introduced by August F. Möbius (1790–1868) in Möbius (1827). Peirce was quite taken with barycentric coordinates, as a mathematical and even as a philosophical device. In a letter draft to James dated 26 February 1909 (Peirce 1897–1909, N3.836–866), Peirce uses them to prove Desargues's Theorem (N3.846–847), and speculates about the utility of an analogous notation in "represent[ing] the modes in which concepts are, or should be, represented as compounded in definitions." For further discussion of this letter, see Eisele (1979b, 282–284); on Möbius's and von Staudt's contributions to projective geometry, see Kline (1972, 850–852).

4. THE SIMPLEST MATHEMATICS

1. The published version of this text, in volume 4 of the *Collected Papers*, was clearly based upon Ms 429, a typescript prepared by his friend

Francis Lathrop's secretary, who seems to have experienced some real difficulties with Peirce's handwriting and with the content of Ms 431, the original. This transcription relies primarily on the latter, preferring Ms 429 only
when it corrects obvious errors in Ms 431 or when it shows clear evidence
(for example, an annotation in Peirce's hand) that he wished to depart from
the longhand manuscript. The typescript appears not to have been proofread
carefully, so subsequent editors may find good reason to reverse some of my
judgments as to the text.

2. *Metaphysics* V.13 (1020a).

3. Proclus (1970, 29).

4. Ammonius (435?–517?) states this view in *In Porphyrii Isagogen*
13.10–11 (Ammonius Hermiae 1891); for Boethius (475?–526?), see *De
Institutione Arithmeticae* (Masi 1983, 72–73).

5. Peirce may be referring to *De Congressu Quaerendae Eruditionis
Gratia* (11, 15–21).

6. *Laws* VII (817e–820e). Peirce's reasons for thinking that this treatment improves on that in the *Republic* are not immediately obvious.

7. The allusion is to Kant's Preface to the Second Edition of the *Critique
of Pure Reason* (B xxx).

8. This ideal of demonstration is derived from the first book of Aristotle's *Posterior Analytics*. Aristotle draws the distinction between "knowledge of the fact" and "knowledge of the reason why" in chapter 13 of that
book.

9. This is a reference, the first of many, to the *intermède* following the
third act of Molière's *Le Malade imaginaire* (1673).

10. Maxwell (1891, Vol. 1, 164).

11. On points at infinity, see note 7 to selection 8 (244). Julius Plücker
(1801–1868) studied inflection points of plane curves in Plücker (1839). The
quaternions are a noncommutative division algebra discovered by William
Rowan Hamilton, who was looking for a number system that could play a
role in three-dimensional space analogous to that of the complex numbers in
two dimensions. Benjamin Peirce was a leading proponent of quaternions in
the United States (Baez 2002, 145–146). See also Kline (1972, 779–782).

12. In the preface to the first edition of Dedekind (1888, 790), Dedekind
speaks of "arithmetic (algebra, analysis) as merely a part of logic."

13. On methodeutic, see selection 1, note 11.

14. Peirce's three normative sciences—esthetics, ethics, and logic—are
the subject of the fifth Harvard lecture on pragmatism (Peirce 1903j); he
concludes his summary of their relations on (200–201) thus: "the morally

good will be the esthetically good specially determined by a peculiar super-added element; and the logically good will be the morally good specially determined by a special superadded element."

15. An "arachnoid film" is a spider web.

16. According to Maxwell's own account he "resolved to read no mathematics on the subject till [he] had first read through Faraday's" work (Maxwell 1891, Vol. 1, 191); the whole passage rather strongly rejects the distinction Peirce draws between Faraday's supposedly non-mathematical approach and Maxwell's own explicitly mathematical one.

17. Uniformitarians held that "the past history of the earth is uniform with the present in terms of the physical laws governing the natural order, the physical processes occuring both within the earth and on its surface, and the general scale and intensity of these processes"; catastrophists, on the other hand, posited catastrophic upheavals and great differences between current and past conditions (Wilson 1973, 417). A leading uniformitarian was the geologist Charles Lyell (1797–1875), a major influence on Darwin.

18. Henry Maudsley (1835–1918), founder of the Maudsley Hospital in South London.

19. Kline (1972, 973–977).

20. At this point the manuscript has the following footnote on divergent series:

> It would not be fair, however, to suppose that every reader will know this. Of course, there are many series so extravagantly divergent that no use at all can be made of them. But even when a series is divergent from the very start, some use might commonly be made of it, if the same information could not otherwise be obtained more easily. The reason is,—or rather, one reason is,—that most series, even when divergent, approximate at last somewhat to geometrical series, at least, for a considerable succession of terms. The series
>
> $$\log(1+x) = x - \frac{1}{2}x^2 + \frac{1}{3}x^3 - \frac{1}{4}x^4 + \text{etc.}$$
>
> is one that would not be judiciously employed in order to find the natural logarithm of 3, which is 1.0986, its successive terms being
>
> $$2 - 2 + \frac{8}{3} - 4 + \frac{32}{5} - \frac{32}{3} + \text{etc.}$$
>
> Still, employing the common device of substituting for the last two terms that are to be used, say M and N, the expression
>
> $$\frac{M}{1-(N/M)}$$

the succession of the first six values is 0.667, 1.143, 1.067, 1.128, 1.067, which do show some approximation to the value. The mean of the last two, which any professional computer would use (supposing him to use this series, at all) would be 1.098, which is not very wrong. Of course, the computer would practically use the series

$$\log 3 = 1 + \frac{1}{12} + \frac{1}{80} + \frac{1}{448} + \text{etc.}$$

of which the terms written give the correct value to four places, if they are properly used.

5. THE ESSENCE OF REASONING

1. Thomson and Tait (1867, 337) say that the impossibility will obtain "until we know thoroughly the nature of matter and the forces which produce its motions."

7. ON THE LOGIC OF QUANTITY

1. See, e.g., Peirce (1908d, E2.443–444). This conviction was surely derived from Peirce's father, who believed that "there was an intimate relation between the structure of the mind—mathematics—and the structure of the universe—ideality. . . . Not only was every part of the physical universe expressible in terms of relatively simple mathematical laws, but every logically consistent mathematical system necessarily had its expression somewhere in nature" (Peterson 1955, 106).

2. Cf. Peirce (1903h, E2.275–283).

3. On Peirce and game-theoretic semantics, see Hilpinen (1983). Pietarinen (forthcoming) suggests a link between Peirce's game-theoretic semantics and the constructivist strain in his philosophy of mathematics.

4. The literal translation of this Latin phrase is 'place in which'. Later in this selection Peirce attributes the phrase to Arthur Cayley (1821–1895), who makes frequent use of it in "A sixth memoir on quantics" (Cayley 1859) to denote the final term of an incidence relation. That is, to say that L is the *locus in quo* of $A_1, A_2, \ldots, A_n$ is to say that the A_i lie in (or on) L. Cayley treats 'space' as an equivalent term, but it should be noted that he allows that a point can be a space in this sense (565), e.g., as the *locus in quo* of the lines in the pencil that it determines (578).

5. Cf. Wittgenstein (1953, paragraphs 143ff) and (Skyrms 2000, 63–65).

6. On association of ideas, see Hume, *Treatise of Human Nature*, 1.1.4.

7. See, e.g., Leibniz (1684, 24–25).

8. SKETCH OF DICHOTOMIC MATHEMATICS

1. See Peirce (1903b, 263–266).

2. Aristotle distinguishes between what (the subsequent tradition called) real and nominal definitions in the second book of the *Posterior Analytics* (see especially 90b4 and 93b29–37): a nominal definition gives the meaning of a term, a real one the essence of a thing.

3. *Posterior Analytics* II.7 (92b19–20).

4. Aristotle defines a postulate in *Posterior Analytics* II.10 (76b) as "what is contrary to the opinion of the learner, which though it is demonstrable is assumed and used without being proved" (tr. Jonathan Barnes). Johan Heiberg (1854–1928) edited what is still the definitive Greek text of the *Elements* (1883–1888).

5. See, e.g., *Philosophia Rationalis sive Logica*, §269.

6. Ernst Schröder (1841–1902) is perhaps best known today for the Schröder-Bernstein Theorem. His major work (Schröder 1880–1905) elaborates Peirce's logic of relatives. Brady (2000) traces this line of development in the algebraic tradition in logic.

7. Desargues's development of projective geometry, in which any two distinct lines intersect, extends the Euclidean plane by adding new points "at infinity" to serve as the points of intersection for parallel lines; these new points taken together make up a line at infinity (Kline 1972, 290). In complex analysis the complex plane is extended by adding a single point at infinity (Stewart and Tall 1983, 206).

8. Peirce is alluding here to Cayley's *Sixth Memoir on Quantics:* see the headnote to selection 16.

9. Peirce gives his topical (topological) postulates for space in a letter of 1896 to W. E. Story, quoted at the beginning of Eisele's introduction to volume II of Peirce (1976).

10. On Aristotle's usage of 'axiom' and 'common notion', see Heath (1926, 120).

11. Peirce has already taken German logicians to task in this selection (60) for their subjectivism; see also Peirce (1903k, E2.251–252).

12. Aristotle says this about the Law of Noncontradiction and the Law of Excluded Middle at *Posterior Analytics* 1.11 (77a10–25).

13. The failure of the exterior angle theorem (*Elements* I.16) is usually traced to the fact that in the elliptic plane lines return upon themselves: see Heath's comments on I.16 (Euclid 1926, 279–281). If Peirce means to connect that with a failure of Euclid's Common Notion 5 ("that the whole is greater than its parts"), it is not clear what the connection is supposed to be. As the proof of the exterior angle theorem in (Hartshorne 2000, 101) shows, the function of Common Notion 5 in Euclid's proof of I.16 is to make up for his lack of explicit axioms of betweenness. Peirce's remark about "triangles whose sides pass through infinity" seems to be a reference to points at infinity in projective geometry, but here again it is not exactly clear what he has in mind.

14. *Elements* I.5 ("In isosceles triangles the angles at the base are equal to one another, and, if the equal straight lines be produced further, the angles under the base will be equal to one another").

15. In Peirce's abortive revision of his father's geometry text, the first part of the section on metrical geometry, entitled "The Philosophy of Metrics," begins with a chapter on "The Principles of Measurement" (Peirce 1894, N2.429–444), which develops the theory of the metron, the freely movable rigid body that is used to define equality of distance.

16. Hintikka (1983, 112) argues that if "rules of inference . . . are restricted to 'natural' ones in a sense that can be easily defined, not all theorems of an undecidable theory can be converted into corollaries."

17. Heath (1926, 129) translates ἔκθεσις as 'setting out': "[it] marks off what is given, by itself, and adapts it beforehand for use in the investigation."

18. See Peirce (1904a).

19. Peirce points here to a well-known logical gap in Euclid's proof of the very first proposition in the *Elements;* it is usually handled nowadays with an explicit axiom ensuring that circles will intersect under appropriate conditions: see, e.g., Hartshorne (2000, 107–108). It is not clear whether Peirce is suggesting here that Euclid's definitions and postulates already suffice to close the gap, or whether they would do so if suitably modified. The third postulate is "To describe a circle with any center and distance" (Euclid 1926, 154).

9. [PRAGMATISM AND MATHEMATICS]

1. Peirce may be thinking of the discussion of geometrical diagrams in *System of Logic* II.iii (Mill 1843, 190–192). The major treatment of geometry in that work is in II.v (224–251). Brady (2000, 21–22) argues that the logic of relatives grew out of Peirce's attempts to represent geometrical reasoning.

2. Karl Weierstrass (1815–1897) is generally credited, along with Cauchy, with the introduction of rigor into the foundations of analysis. "Weierstrass's inheritance of Cauchy's style meant not only improving rigor but also avoiding geometry" (Grattan-Guinness 1997, 482). Dauben (1977, 132) cites a letter draft to Francis Russell, dated 18 September 1908, in which Peirce writes that "the whole Weierstrassian mathematics is characterized by a *distrust of intuition.* Therein it betrays ignorance of a principle of logic of the utmost practical importance; namely that every deductive inference is performed, and can only be performed, by imagining an instance in which the premises are true and *observing* by contemplation of the image that the conclusion is true" (Peirce 1887–1909, N3.968).

3. In 1891 Peirce read a paper at the National Academy of Sciences entitled "Astronomical methods of determining the curvature of space." Though the entire lecture text appears to have been lost, materials related to it will appear in volume 8 (228–230) of the *Writings.* See also Dipert (1977, 411–412).

4. This appears to be a somewhat exaggerated summary of a number of results in complex analysis, to the effect that certain classes of functions are (largely) determined by their zeros and poles.

5. Mill (1843, 215–218).

6. Dedekind (1888) is restated in Schröder (1880–1905, vol. 3, 346–387).

7. "That which any true proposition asserts is real, in the sense of being as it is regardless of what you or I may think about it. Let this proposition be a general conditional proposition as to the future, and it is a real general such as is calculated really to influence human conduct; and such the pragmaticist holds to be the rational purport of every concept" (Peirce 1905c, E2:343).

8. A reference to Jeannot's knife: see the headnote to selection 26 (207).

9. Peirce (1870, W2.360). Peirce defines inclusion as a transitive, antisymmetric and reflexive relation, that is, as what is now known as a partial ordering (Brady 2000, 27)—hence the connection with the $\leq$ relation that generates "the sequence of quantity." The inclusion symbol is "used by

Peirce ambiguously for both inclusion between classes and implication between propositions" (24). On the importance of this ordering for Peirce's axiomatization of the natural numbers, see Shields (1997, 45). Levy (1986) emphasizes the primacy of ordinal conceptions, in particular of total orderings (24–30), in Peirce's analyses of number.

10. Prolegomena to an Apology for Pragmaticism

1. See, e.g., "What Pragmatism Is" (Peirce 1905c, E2:334–345).
2. Friedrich Lange (1828–1875) was a prominent neo-Kantian. Peirce is probably alluding to Lange (1877).
3. For more on Peirce's classifications of signs, see Short (2007, 207–262).

11. ['Collection' in the *Century Dictionary*]

1. Dipert (1997) is a very rich discussion of Peirce's attempts to get clear on the way in which a collection is an individual, and on the ontological connection between a collection and its members.
2. Peirce gives his reasons for preferring 'multitude' to 'cardinal number' in selection 12 (94).

12. [On Collections and Substantive Possibility]

1. On Peirce's reasons for taking this view, see Levy (1997, 100).
2. The "uselessly technical and pedantic" work is not *Principia Mathematica* but rather *Principles of Mathematics* (Russell 1902), which Peirce attributed to both authors on more than one occasion (Hawkins 1997, 115); this indicates, not confusion on Peirce's part, but a suspicion that Russell relied heavily on Whitehead even in works published under his name only. For Schröder's exposition of Dedekind (1888), see note 6 to selection 9 (246). Peirce arguably exaggerates his independence of Cantor in these remarks. As Murphey (1961, 240) notes, Peirce's first reading of Cantor took place around 1884; this was three years after the publication of Peirce (1881), which is probably why Peirce feels justified in saying here that he began his studies of multitude before reading Cantor. It is more of a stretch

to say that he had *completed* his studies before that. Though Peirce appears not to have seen Cantor's diagonal argument (Cantor 1891) for Cantor's Theorem at the time that he found his own (Moore 2007b, 238–243), that discovery does seem to have been stimulated by his reading of Cantor (1895). Given the centrality of this result to Peirce's later theory of infinity, that theory is not so completely independent of Cantor as he tries to let on here.

13. [THE ONTOLOGY OF COLLECTIONS]

1. I.e., 'collection'.

2. "If a sign, B, only signifies characters that are elements (or the whole) of the meaning of another sign, A, then B is said to be a *predicate* (or *essential part*) of A. If a sign, A, only denotes real objects that are a part or the whole of the objects denoted by another sign, B, then A is said to be a *subject* (or *substantial part*) of B. The totality of the predicates of a sign, and also the totality of the characters it signifies, are indifferently each called its logical *depth*. This is the oldest and most convenient term. Synonyms are the *comprehension* of the Port-Royalists, the *content* (*Inhalt*) of the Germans, the *force* of De Morgan, the *connotation* of J. S. Mill. (The last is objectionable.) The totality of the subject, and also, indifferently, the totality of the real objects of a sign, is called the logical *breadth*. This is the oldest and most convenient term. Synonyms are the *extension* of the Port-Royalists (ill-called *extent* by some modern French logicians), the *sphere* (*Umfang*) of translators from the German, the *scope* of De Morgan, the *denotation* of J. S. Mill" (Peirce 1904(?)a, E2.305). The terms 'breadth' and 'depth' are due to Hamilton (1860, 100–104).

3. De Morgan (1863, 333).

4. In an appendix to the Transcendental Dialectic (A642/B670–A704/B732) Kant argues that the Ideas of Reason may legitimately be employed, not as constitutive, but as regulative principles. Paul Guyer and Allen Wood helpfully explain that constitutive principles are "necessary conditions of the possibility of the experience of objects at all" (Guyer and Wood 1998, 18), while the Ideas as regulative principles "are taken to represent not metaphysical beings or entities whose reality is supposed to be demonstrable, but rather goals and directions of inquiry that mark out the ways in which our knowledge is to be sought for and organized" (45).

5. In Chapter VI Russell writes that "a class having only one term is to be identified . . . with that one term" (Russell 1902, 68), a view he then retracts (517–518) in an Appendix dealing with Frege's views.

14. THE LOGIC OF QUANTITY

1. In the wake of Frege's criticisms, we have lost sight of the prestige Mill was accorded in Peirce's day. But as Paul Shields notes (Shields 1997, 44), Frege's critique is itself evidence of Mill's prominence; and Peirce's own landmark treatise on number (Peirce 1881) was designed in part as a refutation of Mill's empiricist philosophy of mathematics. The affinities between Peirce's and Mill's conceptions of mathematics as hypothetical have not yet received the attention they deserve; but see Campos (2007, 473).

2. On Peirce and the analytic/synthetic distinction, see Rosenthal (1984) and Shin (1997). This is a major theme of the literature on corollarial and theorematic reasoning, so many of the works cited in note 23 to the Introduction (p. 231) are relevant here as well.

3. A6/B10–A7/B11.

4. Rudjer Boskovic (1711–1787), Croatian scientist, held that the fundamental constituents of matter were unextended points "distinguished from geometric points only by their possession of inertia and their mutual interaction" (Markovic 1970, 330).

5. "Some A is not some B" is one of the eight basic propositional forms (Hamilton 1860, 529–530) in Hamilton's theory of the quantification of the predicate, his proposed improvement upon the traditional theory of the syllogism. Peirce follows De Morgan (1860a, 257) in reading "Some A is not some B" as "Something that is an A is not identical with something that is a B" (Fogelin 1976, 352).

6. Mill (1865, 71n). The argument is not originally due to Mill, though he defends it; it supposes, not that a third thing could spring up spontaneously from every pair, but that a fifth could spring up from every pair of pairs. The point, as Mill states it, is that "the reverse of the most familiar principles of arithmetic and geometry might have been made conceivable, even to our present mental faculties, if those faculties had coexisted with a totally different consitution of external nature."

7. Kant poses the question at B19.

15. RECREATIONS IN REASONING

1. See especially Hookway (forthcoming).

2. Peirce's own axiomatization in Peirce (1881) was along different lines. A proof of the equivalence of the two sets of axioms is sketched in Shields (1997, 48–49).

3. Charles Godfrey Leland (1824–1903) was a popular authority on the Romani (Gypsies).

4. A slip of the pen: Peirce surely meant to write 'vocable', not 'collection' here.

16. TOPICAL GEOMETRY

1. Havenel (forthcoming) is a very thorough treatment of Peirce's topological work; Dusek (1993) surveys its philosophical ramifications. See also Eisele (1979a) and Murphey (1961, 194–211).

2. See Klein (1924–1928, 105–108) and also Torretti (1978, 139–142).

3. This summary of Cayley and Klein's discoveries is very much indebted to Torretti (1978, 127–129). See Garner (1981, 141–162) for an elegant presentation of the mathematical details.

4. Peirce wrote these subtitles (set here in boldface) in the margins of the manuscript, with a note at the beginning of the manuscript asking the typesetter to do likewise.

5. Let P_1, P_2, P_3, and P_4 be four distinct collinear points in the real projective plane. We can associate with each point P_i a pair of real numbers (x_i, y_i); these *parameters* express that point's position on the line relative to the other three. The cross-ratio of the four points, which is independent of the parameterization, is

$$\frac{(x_1 y_3 - x_3 y_1) \cdot (x_2 y_4 - x_4 y_2)}{(x_2 y_3 - x_3 y_2) \cdot (x_1 y_4 - x_4 y_1)}.$$

The cross-ratio is invariant under projective transformations of the plane Torretti (1978, 120–125). For a more general definition, see Garner (1981, 106).

6. That is, in selection 19.

17. A GEOMETRICO-LOGICAL DISCUSSION

1. Weierstrass (1895). Weierstrass's function, everywhere continuous and nowhere differentiable, is

$$f(x) = \sum_{n=0}^{\infty} b^n \cos(a^n x \pi)$$

where $(0 < b < 1)$, and a is an odd integer such that

$$\frac{2}{3} > \frac{\pi}{ab - 1}.$$

2. "An absurd object such as a round square carries in itself the guarantee of its own non-being in every sense; an ideal object, such as diversity, carries in itself the guarantee of its own non-existence. Anyone who seeks to associate himself with models which have become famous could formulate what has been shown above by saying that the Object as such (without considering the occasional peculiarities or the accompanying Objective-clause which is always present) stands 'beyond being and non-being' [*jenseits von Sein und Nichtsein*]. This may also be expressed in the following less engaging and also less pretentious way, which is in my opinion, however, a more appropriate one: The object is by nature indifferent to being, although at least one of its two Objectives of being, the Object's being or non-being, subsists" (Meinong 1904, 86).

18. ['CONTINUITY' IN THE *CENTURY DICTIONARY*]

1. The ordering of this material in the *Collected Papers* is chronologically misleading. For the correct dating, see Fisch (1971, 246n18) and Havenel (2008).

The classic periodization of Peirce's theories of continuity is given in Potter (1996). Potter identifies a Pre-Cantorian period, ending around 1884, when Peirce first reads Cantor and enters his Cantorian period; the latter gives way to the Kantistic period around 1895. The final, Post-Cantorian, period begins around 1908. The more fine-grained account in Havenel (2008) comprises five periods: Anti-nominalistic (1868–1884), Cantorian (1884–1892), Infinitesimal (1892–1897), Supermultitudinous (1897–1907), and Topological (1908–1913).

2. *Physics* V.3 (227a10ff) and *Metaphysics* XI.11 (1069a5).

3. *Critique of Pure Reason* (A169/B211).

4. Cantor (1883, 903–906). Peirce's definition of a perfect set is different from Cantor's. If we call the point p a *limit point* of the point set S just in case every neighborhood of p contains infinitely many points of S, then according to Peirce's definition S is perfect just in case it contains all of its limit points; Cantor, on the other hand, defines a perfect as one that contains all *and only* its limit points.

5. The first two examples may derive from Cantor (1872, 98); the second is a variation on the ternary set Cantor defines in Cantor (1883, 919).

6. Here Peirce acknowledges his omission of the "only" half of Cantor's definition of a perfect set. The omitted condition is not quite so otiose as Peirce suggests: a point set with only one member is trivially concatenated, but that member is not a limit point of the set. If we rule out such trivial cases by stipulating that a concatenated set contains at least two distinct points, then Peirce's observation is correct. A non-trivial concatenated set consists only of limit points, because we can lay down an arbitrarily fine mesh of points between any two. Peirce may have been taking this stipulation for granted when he wrote this note. Peirce's struggles with the concept of perfect set are reviewed in Moore (forthcoming 2).

7. Peirce criticizes Cantor on this score in selections 19 (151) and 21 (160). The criticism is unfair: as Myrvold (1995, 521) observes, Cantor states quite clearly that he is defining a continuous point set within the n-dimensional manifold R^n, that is, the n-fold Cartesian product of the set R of real numbers with itself (Cantor 1883, 904).

8. This assumption—that the cardinality of the real numbers is the first uncountable cardinal—is Cantor's Continuum Hypothesis: see selection 22, note 1 below (p. 255).

19. THE LAW OF MIND

1. In calling Cantor's and Dedekind's definitions equivalent, I mean that they define isomorphic structures. For further comparative discussion of these definitions and Peirce's, see Myrvold (1995, 522–524) and also Moore (2007a, 465n24).

2. Cantor's rejection of infinitesimals is detailed in Dauben (1990, pp. 128–132, 233–238). As Myrvold (1995, 525) points out, Peirce did not have access to Cantor's more forthright condemnations of the infinitely small when he wrote "The Law of Mind."

3. See, e.g., Peirce (1869, 68–69).

4. This list of journals matches perfectly with the original places of publication of the early papers of Cantor reprinted, in French translation, in volume 2 of *Acta Mathematica*. A copy of that issue of the journal is among Peirce's papers at the Houghton Library (MS CSP 1599, Box 1), with marginal annotations by Peirce, in handwriting that dates them around 1890. Only three of the papers (Cantor 1874, 1878, 1883) are actually annotated; these are also, as it happens, the only Cantorian works which leave unmistakeable traces in "The Law of Mind."

There has been a good deal of comparative discussion of Cantor and Peirce. As a start, see Dauben (1977), Dipert (1997), and Murphey (1961, 274–286). Dipert also considers (58–62) the Boolean influence on Peirce's thinking about collections; for further comparative analysis, see Grattan-Guinness (1997, 29–30). Moore (forthcoming 3) attempts to determine the chronology of Peirce's reading of Cantor's works.

5. See Peirce (1881, W4.309) and De Morgan (1860b, 242–246).

6. This remark is most straightforwardly read as asserting the existence of only one uncountable cardinal. Had Peirce read Cantor (1883) more thoroughly, he would have realized that Cantor had already shown there that there is more than one (911). It is equally evident that as of 1892 Peirce did not know Cantor (1891), in which Cantor proves what is now known as Cantor's Theorem and notes its obvious corollary (922), that every infinite cardinal has a successor. Peirce soon came to have doubts on the matter—see Peirce (1893c, C4.121)—even before his own discovery of Cantor's Theorem a few years later.

7. Peirce (1881, W4.300–301).

8. Pierre de Fermat (1601–1665) is a giant in the history of number theory and of probability theory. In one of his discussions of the "Fermatian inference," Peirce (1893c, C4.110) quotes from a letter (ca. 1659) to Pierre de Carcavi (1600–1684), in which Fermat explains his method of infinite descent (Fermat 1891–1912, Vol. 2, 431–436). In the omitted paragraphs that follow, Peirce illustrates the Fermatian inference with an inductive proof of the binominal theorem.

9. In Cantor (1874, 840–841) Cantor shows that the set of real algebraic numbers is countable. Peirce may think that his result is more general because the "definite rule" F could associate, with each finite array of integers, a polynomial determining an algebraic number, in such a way that a polynomial for every such number would eventually crop up in the enumeration.

10. See Cantor (1874).

11. Cantor proves that the *n*-fold Cartesian product over the real numbers has the same cardinality as the real numbers themselves in Cantor (1878). The argument Peirce gives here is essentially the one that first led Cantor to the result; its flaws were quickly pointed out to him by Dedekind, with whom he worked closely on the corrected (and much less straightforward) proof he eventually published (Dauben 1990, 54–58).

12. Peirce's examples illustrate addition and multiplication of transfinite ordinals (Cantor 1883, 886). For definiteness in the first example, consider the natural number sequence and its ordinal number ω. Then Peirce's construction tacks a copy of the sequence onto the end of it; the result has the ordinal number $\omega + \omega$. In the second example let U be an uncountable set well-ordered by the binary relation $<$. Peirce's construction begins with the Cartesian product U^2 of U with itself, that is, the set of all ordered pairs of elements of U. Well-order that set lexicographically by stipulating that (a_1,b_1) is to be less than (a_2,b_2) if $a_1 < a_2$ or if $a_1 = a_2$ and $b_1 < b_2$. Then if the ordinal number of U as ordered by $<$ is α, the ordinal number of U^2 under the ordering just defined will be $\alpha \cdot \alpha$.

13. This is another variation on Cantor's ternary set (Cantor 1883, 919).

14. Chessin (1896) defines a regular sequence of rational numbers as one "which may be indefinitely extended according to a determinate law

$$\gamma_1, \gamma_2, \gamma_3, \ldots, \gamma_m, \ldots$$

[such that] all its terms remain in absolute amount less than a finite determinate number, and . . . at the same time a number m can be found such that the difference

$$|\gamma_{m+n} - \gamma_m| \ (n = 1,2,3,\ldots)$$

can be made less than an arbitrarily small number ε." That is, a regular sequence is a bounded Cauchy sequence. Since Cauchy sequences are bounded in any case, a regular sequence is just a Cauchy sequence of rationals. A positive answer to Peirce's query is an immediate consequence of Cantor's definition of the real numbers in Cantor (1872). Though this paper was in the issue of *Acta Mathematica* discussed in note 4, it bears no annotations in Peirce's hand, so he was probably not aware of Cantor's definition when he wrote this essay.

15. The simplistic view of Cauchy as the eradicator, with Weierstrass, of the infinitely small from the foundations of analysis was first challenged by Abraham Robinson (1966, 269–276); see Schubring (2005, 427–480) for a recent discussion and survey of the literature. In his influential *Éléments de*

calcul infinitesimal (Duhamel 1856), Jean-Marie Duhamel (1797–1872) employs a principle licensing the substitution of quantities whose differences are infinitely small; again, see Schubring (2005, 587–598).

20. [SCIENTIFIC FALLIBILISM]

1. The quote is from Act I of W. S. Gilbert's *Engaged* (1877). Peirce has misremembered the name of the character (Belinda, not Matilda) who speaks these lines: "Belvawney, I love you with an imperishable ardour which mocks the power of words. If I were to begin to tell you now of the force of my indomitable passion for you, the tomb would close over me before I could exhaust the entrancing subject."

2. Ludwig Büchner (1824–1899), German physician and philosopher. His *Kraft und Stoff* (Frankfurt: Meidinger, 1855) was a seminal work of nineteenth century German materialism.

22. THE DISPUTE BETWEEN NOMINALISTS AND REALISTS

1. The Continuum Hypothesis says that the cardinality of the real numbers, which is identical to that of the power set (set of all subsets) of the natural numbers, is the next cardinal greater than that of the natural numbers. The Generalized Continuum Hypothesis says that for *any* infinite cardinal κ, the next cardinal after κ is the cardinal number of the power set of a set with cardinality κ. Both are now known to be independent of the generally accepted axioms of set theory. Cantor devoted a great deal of energy to the futile attempt to prove the weaker hypothesis; Peirce seems to have taken the truth of the stronger for granted (Myrvold 1995, 515). For a nice overview of these and other independent questions of set theory, see Maddy (1997, 63–72).

2. In his announcement of the classification (Peirce 1866, W1.428), Peirce uses 'hypothesis' rather than 'retroduction'. See also Peirce (1867).

3. Cf. selection 1, note 11 (p. 238).

4. In Peirce (1867).

5. A simple example of the sort of thing Peirce has in mind would be the transition from *A touches B* to *A is next to B*. (Special thanks to André De Tienne for suggesting this interpretation to me, and for assistance in formulating this note.)

6. Here Peirce is clearly assuming the Generalized Continuum Hypothesis (see note 1, this selection) in his account of the transition from one infinite multitude to the next. The further claim, that there only countably many of these infinite multitudes, rests on an argument which Peirce spells out in full in selection 24 (191).

7. Cantor (1895, 89–90).

8. Hilary Putnam (1992, 41–45) argues that in this passage Peirce adopts a view of points with echoes in nonstandard analysis, where every point on the line is surrounded by a "monad" of points lying infinitely close to it. See also Moore (2007a, 448–450) and the works cited there.

9. Abbot (1885, 1–5).

10. Cf. Peirce (1877, E1.120), where "the method of science" is said to have as "its fundamental hypothesis . . . [that] there are real things, whose characters are entirely independent of our opinions about them."

11. This interpretation of evolution is developed at length in Peirce (1893b). Short (2007, 117–150) is a magnificent overview of Peirce's views on final causation, and its role in evolution.

23. The Logic of Continuity

1. See Havenel (forthcoming) for a detailed and lucid treatment of Peirce's analysis of singularities in the context of the Census Theorem. Murphey (1961, 194–211) and the comments on this lecture in Putnam and Ketner's edition (R.99–101) are also valuable resources.

2. The Paradox is usually taken to show that there can be no universal set U, that is, no set containing everything. For every subset of U would be an element of U, and so its power set $P(U)$ would be a subset of U. Therefore in particular $P(U)$ cannot have more members than U; but this violates Cantor's Theorem, which says that a set always has fewer members than its power set. Peirce proves his version of Cantor's Theorem in selection 22; for Cantor's formulation of the Paradox, see his letter to Dedekind of 31 August 1899, in Ewald (1996, 939–940). Myrvold (1995) is a valuable treatment of Peirce's version of Cantor's Paradox, and its role in his theory of continuity; see also Murphey (1961, 259–267).

3. Peirce's approach to proper classes is compared with Cantor's in Dipert (1997, 66–67).

4. The paper Peirce refers to is "The Law of Mind."

5. Putnam and Ketner insert '1890' into this blank in the manuscript (R.242). Though there is no way of knowing for certain what date Peirce ultimately decided upon, this seems late in view of Peirce's enumeration, in selection 19 (147), of the "mathematical journals of the first distinction" in which Cantor's writings had appeared before being published together, in French translation, in *Acta Mathematica* in 1883.

6. Peirce left a blank in the manuscript here.

7. Gaspard Monge (1746–1818) also did important work in differential geometry. His descriptive geometry built on techniques of projection first developed by Albrecht Dürer (1471–1528) (Kline 1972, 235).

8. Clifford (1886, 80).

9. Putnam and Ketner (R.246) insert '1859' into this blank in the manuscript. But why did Peirce leave it blank, as if he needed to check the date before filling it in, when he had confidently written the date just a few lines before? Moreover, the repetition of the date that results from Putnam and Ketner's interpolation is rather awkward. More likely, perhaps, that Peirce's intention was to fill in the title or place of publication for Cayley's memoir.

10. Johann Benedikt Listing (1808–1882) proved the Census Theorem, which Peirce devoted a great deal of energy to elaborating (see selection 16n1). Peter Guthrie Tait (1831–1901) came to Listing's topological work through his own interest in knots (Przytycki 1998, 537). Listing's memoirs are Listing (1847) and (1862). In the paper to which Peirce alludes (Riemann 1851), Riemann analyzes the connectivity of the surfaces that bear his name (see selection 1, note 2). "The greatest impetus to topological investigations came from Riemann's work in complex function theory" (Kline 1972, 1166).

24. [ON MULTITUDES]

1. The manuscript uses the Aries symbol ('♈') to denote the "denumeral multitude."

2. The fallacy in this argument is discussed in Murphey (1961, 262), where John Myhill is credited with its discovery, and also in Myrvold (1995, 517). The mis-step is the assertion that a countably infinite sequence is unaltered by the addition of a new term. This is correct if we add the new term at the beginning of the sequence; for example, when we put a new number before 0 in the natural number sequence, we get a sequence with the same order type as the natural numbers themselves. But if we add the new term *after* all the natural numbers, the order type is changed. Peirce's argument

supposes that we are adding a new term at the bottom of his infinite tower of exponentiations, when in fact we are adding it at the top.

3. Peirce left a blank in the manuscript here, presumably to be filled in when the memoir was complete.

4. This is clearly a slip of the pen: 108 is obviously what is wanted here.

5. The primipostnumeral multitude is what we would now call the first uncountable cardinal, the secundipostnumeral the second, and so on. Because Peirce accepts the Generalized Continuum Hypothesis, he differs from most set theorists today on the identities of these numbers.

6. Simon Newcomb, *Elements of the Differential and Integral Calculus* (New York: Henry Holt and Co., 1887).

7. *Elements* X.1, the basis of the method of exhaustion: "Two unequal magnitudes being set out, if from the greater there be subtracted a magnitude greater than its half, and from that which is left a magnitude greater than its half, and if this process be repeated continually, there will be left some magnitude which will be less than the lesser magnitude set out" (Euclid 1926, 14).

25. Infinitesimals

1. In the ensuing months Peirce made a concerted effort to obtain Cantor's more recent writings, which he finally succeeded in doing near the end of 1900. In December of that year he composed two letter drafts to Cantor in which he attempts to address many of the questions about collections that would preoccupy him in the years to come. The sequence of events is discussed in more detail in Moore (forthcoming 3). The philosophical content of these letter drafts is explored in Dipert (1997, 68–70).

2. The "small books by Dedekind" are Dedekind (1872) and (1888); the "memoirs by Cantor" are the two parts of the *Beiträge* (Cantor 1895; 1897); and "Schröder's *Logic*" is Schröder (1880–1905). The assertion of Peirce's priority, or at least independence, in much of the mathematical content of this letter is a poorly hidden agenda here.

3. Peirce (1881, W4.309).

4. "A system S is said to be *infinite* when it is similar to a proper part [*echten Theile*] of itself; otherwise S is said to be a finite system" (Dedekind 1888, 64). That is, a set is (Dedekind-)infinite just in case one of its proper subsets can be mapped onto it one to one. For comparative discussion of Peirce's and Dedekind's definitions of finiteness, see Moore (1982, 25) and

Shields (1997, 49–50). Randall Dipert (1997, 62–64) gives a philosophical reconstruction of Peirce's argument for the equivalence: in order to know that the Syllogism of Transposed Quantity fails (and hence that the set is infinite) we must possess detailed information about certain relations on the set, and such information is also required if we are to know that a proper subset of an infinite set *is* a proper subset. Dipert points out (74n23) that Peirce errs in his account of Dedekind's definition: the correct requirement is that the system be similar to *some* (not every) proper part of itself.

5. Royce (1899, 562n2). Royce mentions Italy because of the work of the Giuseppe Veronese (1854–1917), whose use of infinitesimals in geometry particularly incensed Cantor (Dauben 1990, 233–238).

6. Peirce (1897, C3.548). The argument Peirce gives there is the diagonal argument he gives in selection 22 (171), with some sugar coating.

7. Royce (1899, 562n1). The passage Royce refers to is selection 19 (147). He offers quite decisive evidence against Peirce's reading of Cantor from the latter's own works, including his argument for the impossibility of infinitesimal linear magnitudes (Cantor 1887–1888, 407–409).

8. Cantor (1885, 275–276). Cantor would have violently rejected the suggestion that it follows from his hypothesis that "infinitesimals must be actual real distances." Peirce thought Cantor's proposal could cast light on the interaction between mind and matter: see Dauben (1977, 130–131).

26. THE BEDROCK BENEATH PRAGMATICISM

1. Alfred Bray Kempe (1849–1922) was an accomplished amateur mathematician, originator of an important failed proof of the Four Color Theorem. Peirce was a great, though not uncritical, admirer of Kempe's *Memoir on the Theory of Mathematical Form* (Kempe 1886), the "marvellously strong contribution" he praises here: see especially Peirce (1903a, E2.173–176), on Kempe's analysis of mathematical diagrams. It is not clear why the *Memoir* led Peirce to think that the concept of collection is "indecomposable." (Kempe explains his usage of 'collection' on p. 3.) For more on Peirce and Kempe, see Grattan-Guinness (1997, 35–36; 2002).

27. Note and Addendum on Continuity

1. A systematic study of Choice-related principles and Peirce's writings on collections is urgently needed. Randall Dipert (1997, 84) argues that in his critical commentary on Dedekind's definition of infinite collections (see note 4 to selection 25, p. 258 above), "Peirce is dealing with issues related to choice functions and the Axiom of Choice." Dipert says more about Choice on pp. 63–65 of his paper; see also Myrvold (1995, 529–535).

2. Cantor defines ω^ω as the least ordinal greater than all ordinals of the form

$$\nu_0\omega^\mu + \nu_1\omega^{\mu-1} + \ldots + \nu_{\mu-1}\omega + \nu_\mu$$

"where $\mu, \nu_0, \nu_1, \ldots, \nu_m$ assume all finite, positive integral values including zero and excluding the case where $\nu_0 = \nu_1 = \ldots = \nu_\mu = 0$ " (Cantor 1883, 909). The bearing of Peirce's discussion on ω^ω is quite clear from that passage, so Peirce is probably not defining ω^ω here as Cantor does. In a draft of a letter to Francis Russell, dated 18 September 1908, Peirce says that ω^ω and 2^ω are "the same multitude" (Peirce 1887–1909, 973) so he is probably failing here, as he does there, to distinguish as Cantor does between ordinal and cardinal exponentiation; in that case the force of this parenthetical remark is that Peirce is uncertain about how the ordinals he discusses in the text bear on what he believed to be the first abnumeral multitude. See Murphey (1961, 267–269) for more on this confusion in Peirce's reading of Cantor, and on this letter to Russell.

28. Addition [on Continuity]

1. See especially Peirce's correspondence with Josiah Royce and E. H. Moore in the third volume of Peirce (1976). Some of his other attempts to prove Trichotomy, and Peirce's doubts about his own arguments, are discussed in Moore (1982, 49). Moore also identifies the equivalents of Choice at work in Peirce's proofs.

2. For an account of Borel's publication of Bernstein's proof, and his own attitude towards Trichotomy, see Moore (1982, 49–50). Borel's opposition to the Axiom of Choice is forcefully stated in his famous correspondence with René Baire (1874–1932), Henri Lebesgue (1875–1941) and Jacques Hadamard (1865–1963). These letters are reproduced in an appendix (311–320) to Moore's book.

3. Arthur Schönflies (1853–1928), for example, believed that Trichotomy was weaker than Well-Ordering (Moore 1982, 117). It was not until 1915 that Friedrich Hartogs (1874–1943) proved that the Axiom of Choice implies Trichotomy (170). Philip Jourdain (1879–1919) stated this implication without proof in 1907 (133). Peirce refers to Borel's supposed proof in his letter draft to Jourdain (Peirce 1908c, N3.880), dated 5 December 1908, so his own hunch about Trichotomy and Well-Ordering is probably uninfluenced by Jourdain.

4. This is a reference to the concluding section (C3.526–552) of Peirce (1897).

29. SUPPLEMENT [ON CONTINUITY]

1. For a sympathetic discussion of this last phase of Peirce's thinking about continuity, see Havenel (2008, 117–125).

2. Riemann (1867).

3. Cf. Peirce (1871, E1.88): "The real is that which is not whatever we happen to think it, but is unaffected by what we may think of it."

Bibliography

Note on page references: Because many of the works cited in this book are most readily available in editions other than the first, I have adopted the following convention whenever it will save space, while facilitating the use of original dates of publication in citations. When page references in the text are taken from a later edition or reprinting, I give a citation to the later source in square brackets at the end of the entry for the original publication. For Peirce's works these bracketed citations will use the codes for previous collections of his works, as explained in the preface (p. ix).

1. WRITINGS OF PEIRCE

1866. *The Logic of Science* [Lowell Lectures of 1866]. Mss 343, 345, 352–359, 696, 919, 1571. [W1]

1867. On a new list of categories. *Proceedings of the American Academy of Arts and Sciences* 7, 287–298. [E1]

1869. Grounds of validity of the laws of logic. *Journal of Speculative Philosophy* 2, 193–208. [E1]

1870. Description of a notation for the logic of relatives, resulting from an amplification of the conception of Boole's calculus of logic. *Memoirs of the American Academy of Arts and Sciences* n.s. 9, 317–378. [W2]

1871. Fraser's *The Works of George Berkeley. North American Review* 113 (October), 449–472. [E1]

1877. The fixation of belief. *Popular Science Monthly* 12 (November), 1–15. [E1]

1878. How to make our ideas clear. *Popular Science Monthly* 12 (January), 286–302. [E1]

1881. On the logic of number. *American Journal of Mathematics* 4, 85–95. [W4]

1885. An American Plato. Ms 1369. [E1]

1887–1909. Letters to Francis Russell. Ms L387. [N3]

1888(?)–1914(?) [Critical annotations in Peirce's copy of the *Century Dictionary*]. MS CSP 1597, Houghton Library, Harvard University.

1892. The law of mind. *Monist* 2, 533–559. [E1]

1893a. The essence of reasoning [*Grand Logic,* Division III, Chapter VI]. MS CSP 409, Houghton Library, Harvard University.

1893b. Evolutionary love. *Monist* 3, 176–200. [E1]

1893c. The logic of quantity [*Grand Logic,* Book III, Chapter 17]. MS CSP 423 and MS CSP 1009, Houghton Library, Harvard University. [C4]

1893d. [Scientific fallibilism.] MS CSP 955, Houghton Library, Harvard University.

1894. New Elements of Geometry by Benjamin Peirce, Rewritten by his Sons, James Mills Peirce and Charles Sanders Peirce. MS CSP 94, Houghton Library, Harvard University. [N2]

1895(?)a. Elements of Mathematics. Ms 165. [N3]

1895(?)b. On quantity, with special reference to collectional and mathematical infinity. MS CSP 15, Houghton Library, Harvard University.

1895(?)c. On the logic of quantity, and especially of infinity. MS CSP 16, Houghton Library, Harvard University.

1896. The regenerated logic. *Monist* 7, 19–40. [C3]

1896(?). On quantity, with special reference to collectional and mathematical infinity. MS CSP 14, Houghton Library, Harvard University.

1897. The Logic of Relatives. *Monist* 7, 161–217. [C3]

1897(?)a. [On multitudes]. MS CSP 28, Houghton Library, Harvard University. [N3]

1897(?)b. Recreations in reasoning. MS CSP 205, Houghton Library, Harvard University. [C4]

1897–1909. Letters to William James. Ms L224. [N3]

1898a. Detached ideas continued and the dispute between nominalists and realists. MS CSP 439, Houghton Library, Harvard University. [R]

1898b. The logic of continuity. MS CSP 948, Houghton Library, Harvard University. [R]

1898c. The logic of mathematics in relation to education. *Educational Review* 15(3), 209–216. [C3]

1900. Infinitesimals. *Science* ns 11, 430–433 (March 16). [C3]

1901. Definitions for Baldwin's *Dictionary.* In J. M. Baldwin, ed., *Dictionary of Philosophy and Psychology.* New York: Macmillan.

1902. The simplest mathematics. [*Minute Logic,* Chapter 3]. MS CSP 431, Houghton Library, Harvard University, with corrections from MS CSP 429.

1903a. The categories defended [Harvard Lectures on Pragmatism, Lecture 3]. Ms 308. [E2]

1903b. The ethics of terminology. Ms 478. [E2]

1903c. Lecture II [Rejected draft, Harvard Lectures on Pragmatism]. MS CSP 303, Houghton Library, Harvard University. [N4]

1903d. [Lowell Lecture 5]. MS CSP 469, Houghton Library, Harvard University. [N3]

1903e. The maxim of pragmatism [Harvard Lectures on Pragmatism, Lecture 1]. Ms 301. [E2]

1903f. The nature of meaning [Harvard Lectures on Pragmatism, Lecture 6]. Mss 314, 316. [E2]

1903g. [Rejected draft, Lowell Lecture 3]. MS CSP 459, Houghton Library, Harvard University. [N3]

1903h. Sundry logical conceptions. Ms 478. [E2]

1903i. *A Syllabus of Certain Topics of Logic.* Boston: Alfred Mudge and Son. [E2]

1903j. The three normative sciences [Harvard Lectures on Pragmatism, Lecture 5]. Ms 312. [E2]

1903k. What makes a reasoning sound? Mss 448–449. [E2]

1903(?). Sketch of dichotomic mathematics. MS CSP 4, Houghton Library, Harvard University.

1904a. Ideas, stray or stolen, about Scientific Writing. Ms 774. [E2]

1904b. Letter to Lady Welby (12 October). [C8]

1904(?)a. New elements. Ms 517 [E2].

1904(?)b. Topical Geometry. MS CSP 137, Houghton Library, Harvard University.

1905a. The bed-rock beneath pragmaticism. MS CSP 300, Houghton Library, Harvard University. [C6]

1905b. Issues of pragmaticism. *Monist* 15, 481–499. [E2]

1905c. What pragmatism is. *Monist* 15, 161–181. [E2]

1906. Prolegomena to an apology for pragmaticism. *Monist* 16, 492–546.

1906(?). A geometrico-logical discussion. MS CSP 126, Houghton Library, Harvard University.

1908a. Addition. MS CSP 203, Houghton Library, Harvard University.

1908b. Letter to Lady Welby (23 December). Ms L463. [E2]

1908c. Letter to Philip Jourdain (5 December). Ms L230a. [N3]

1908d. A neglected argument for the reality of God. *Hibbert Journal* 7 (October), 90–112. [E2]

1908e. Some amazing mazes (conclusion): Explanation of curiosity the first. *Monist* 18, 416–464.

1908f. Supplement. MS CSP 204, Houghton Library, Harvard University.

1931–1958. *The Collected Papers of Charles Sanders Peirce.* 8 vols. Cambridge: Harvard University Press. Edited by Charles Hartshorne, Paul Weiss and Arthur Burks.

1976. *The New Elements of Mathematics.* 4 vols. Edited by Carolyn Eisele. The Hague: Mouton.

1992. *Reasoning and the Logic of Things.* Cambridge: Harvard University Press. Edited by Kenneth Laine Ketner, with introduction by Kenneth Laine Ketner and Hilary Putnam.

1998. *The Essential Peirce 2 (1867–1893).* Bloomington: Indiana University Press. Edited by the Peirce Edition Project.

2. Secondary Works on Peirce's Philosophy of Mathematics

Alborn, Timothy L. 1989. Peirce's evolutionary logic: Continuity, indeterminacy, and the natural order. *Transactions of the Charles S. Peirce Society* 25, 1–28.

Campos, Daniel G. 2007. Peirce on the role of poietic creation in mathematical reasoning. *Transactions of the Charles S. Peirce Society* 43, 470–489.

———. 2009. Imagination, concentration and generalization: Peirce on the reasoning abilities of the mathematician. *Transactions of the Charles S. Peirce Society* 45, 135–156,

———. Forthcoming. The imagination and hypothesis-making in mathematics: A Peircean account. In Moore (forthcoming 1).

Cooke, Elizabeth F. 2003. Peirce, fallibilism, and the science of mathematics. *Philosophia Mathematica (III)* 11, 158–175.

———. Peirce's general theory of inquiry and the problem of mathematics. In Moore (forthcoming 1).

Dauben, Joseph W. 1977. C. S. Peirce's philosophy of infinite sets. *Mathematics Magazine* 50, 123–135.

———. 1981. Peirce on continuity and his critique of Cantor and Dedekind. In Kenneth L. Ketner, Joseph M. Ransdell, Carolyn Eisele, Max H. Fisch, and Charles S. Hardwick, eds., *Graduate Studies Texas Tech University,* 93–98. Lubbock: Texas Tech Press.

———. 1982. Peirce's place in mathematics. *Historia Mathematica* 9, 311–325.

———. 1995. Peirce and history of science. In Ketner (1995), 146–195.

———. 1996. Charles S. Peirce, evolutionary pragmatism and the history of science. *Cenaturus* 38, 22–82.

Dea, Shannon. 2006. "Merely a veil over the living thought": Mathematics and logic in Peirce's forgotten Spinoza review. *Transactions of the Charles S. Peirce Society* 42, 501–517.

de Waal, Cornelis. 2005. Why metaphysics needs logic, and mathematics doesn't: Mathematics, logic and metaphysics in Peirce's classification of the sciences. *Transactions of the Charles S. Peirce Society* 41, 283–297.

Dipert, Randall R. 1977. Peirce's theory of the geometrical structure of physical space. *Isis* 68, 404–413.

———. 1997. Peirce's philosophical conception of sets. In Houser, Roberts, and Van Evra (1997), 53–76.

Dusek, R. Valentine. 1993. Peirce as philosophical topologist. In Moore (1993), 49–59.

Ehrlich, Philip. Forthcoming. The absolute arithmetic continuum and its Peircean counterpart. In Moore (forthcoming 1).

Eisele, Carolyn. 1964. Peirce's philosophy of mathematical education. In Edward C. Moore and Richard Robin, eds., *Studies in the Philosophy of Charles Sanders Peirce, Second Series,* 51–75. Amherst: University of Massachusetts Press. [Eisele 1979g, 177–200]

———. 1979a. The four-color problem. In Eisele (1979g), 216–222.

———. 1979b. Mathematical exactitude in the doctrine of exact philosophy. In Eisele (1979g), 276–291.

———. 1979c. The mathematical foundations of Peirce's philosophy. In Eisele (1979g), 237–244.

———. 1979d. Peirce as a precursor in mathematics and science. In Eisele (1979g), 292–299.

———. 1979e. The problem of mathematical continuity. In Eisele (1979g), 208–215.

———. 1979f. The role of modern geometry in Peirce's philosophical thought. In Eisele (1979g), 245–250.

———. 1979g. *Studies in the Scientific and Mathematical Philosophy of Charles S. Peirce.* Edited by R. M. Martin. The Hague: Mouton.

———. 1995. Charles S. Peirce as mathematician. In Ketner (1995), 120–131.

———. 1996. The Modern Relevance of the Mathematical Philosophy of Charles S. Peirce. In Robert W. Burch, ed., *Frontiers in American Philosophy Volume II,* 79–85. College Station: Texas A&M Univ Pr.

Fisch, Max H. 1981. Peirce as scientist, mathematician, historian, logician and philosopher. In Kenneth L. Ketner et al., eds., *Proceedings of the C. S. Peirce Bicentennial International Congress*, 13–34. Lubbock: Texas Tech Press. [Fisch 1986, 376–400]

Grattan-Guinness, Ivor. 1997. Peirce between logic and mathematics. In Houser, Roberts, and Van Evra (1997), 23–42.

———. 2002. Kempe on multisets and Peirce on graphs, 1886–1905. *Transactions of the Charles S. Peirce Society* 38, 327–350.

Haack, Susan. 1979. Fallibilism and necessity. *Synthese* 41, 37–64.

———. 1993. Peirce and logicism: Notes towards an exposition. *Transactions of the Charles S. Peirce Society* 29, 34–56.

Havenel, Jérôme. 2008. Peirce's clarifications of continuity. *Transactions of the Charles S. Peirce Society* 44, 86–133.

———. Forthcoming. Peirce's topological concepts. In Moore (forthcoming 1).

Hawkins, Benjamin S., Jr. 1997. Peirce and Russell: The history of a neglected "controversy." In Houser, Roberts, and Van Evra (1997), 111–146.

Herron, Timothy. 1997. C. S. Peirce's theories of infinitesimals. *Transactions of the Charles S. Peirce Society* 33, 590–645.

Hintikka, Jaakko. 1983. C. S. Peirce's "first real discovery" and its contemporary relevance. In Freeman (1983), 107–118.

Hoffmann, Michael H. G. 2003. Peirce's "diagrammatic reasoning" as a solution to the learning paradox. In Guy Debrock, ed., *Process Pragmatism: Essays on a Quiet Philosophical Revolution*, 121–143. Amsterdam: Rodopi.

Hookway, Christopher. Forthcoming. "The form of a relation": Peirce and mathematical structuralism. In Moore (forthcoming 1).

Houser, Nathan. 1993. On "Peirce and logicism": A response to Haack. *Transactions of the Charles S. Peirce Society* 29, 57–67.

Hull, Kathleen. 1994. Why hanker after logic? Mathematical imagination, creativity and perception in Peirce's systematic philosophy. *Transactions of the Charles S. Peirce Society* 30, 271–295.

———. 2005. The inner chambers of his mind: Peirce's "Neglected argument" for God as related to mathematical experience. *Transactions of the Charles S. Peirce Society* 41, 484–513.

Johanson, Arnold. 2001. Modern topology and Peirce's theory of the continuum. *Transactions of the Charles S. Peirce Society* 37, 1–12.

Ketner, Kenneth Laine. 1985. How Hintikka misunderstood Peirce's account of theorematic reasoning. *Transactions of the Charles S. Peirce Society* 21, 407–418.

Kerr-Lawson, Angus. 1997. Peirce's pre-logistic account of mathematics. In Houser, Roberts, and Van Evra (1997), 77–84.

Lenhard, Johannes. 2004. Scepticism and mathematization: Pascal and Peirce on mathematical epistemology. *Philosophica* 74, 85–102.

Levy, Stephen H. 1986. Peirce's ordinal conception of number. *Transactions of the Charles S. Peirce Society* 22, 23–42.

———. 1991. Charles S. Peirce's theory of infinitesimals. *International Philosophical Quarterly* 31, 127–140.

———. 1997. Peirce's theoremic/corollarial distinction and the interconnections between mathematics and logic. In Houser, Roberts, and Van Evra (1997), 85–110.

Locke, Gordon. 2000. Peirce's metaphysics: Evolution, synechism, and the mathematical conception of the continuum. *Transactions of the Charles S. Peirce Society* 36, 133–147.

Marietti, Susanna. 2005. Mathematical individuality in Charles Sanders Peirce. *Cognitio* (July–December), 201–207.

———. 2006. Semiotics and Deduction: Perceptual Representations of Mathematical Processes. In S. Marietti and R. Fabbrichesi, eds., *Semiotics and Philosophy in Charles Sanders Peirce,* 112–127. Cambridge: Cambridge Scholars Press.

———. Forthcoming. Observing signs. In Moore (forthcoming 1).

Moore, Matthew E. 2007a. The genesis of the Peircean continuum. *Transactions of the Charles S. Peirce Society* 43, 425–469.

———. 2007b. On Peirce's discovery of Cantor's Theorem. *Cognitio* 8, 223–248.

———, ed. Forthcoming 1. *New Essays on Peirce's Mathematical Philosophy.* Chicago: Open Court.

———. Forthcoming 2. Peirce on perfect sets, revised. Forthcoming in *Transactions of the Charles S. Peirce Society.*

———. Forthcoming 3. Peirce's Cantor. In Moore (forthcoming 1).

Myrvold, Wayne C. 1995. Peirce on Cantor's Paradox and the continuum. *Transactions of the Charles S. Peirce Society* 31, 509–541.

Noble, N. A. Brian. 1989. Peirce's definitions of continuity and the concept of possibility. *Transactions of the Charles S. Peirce Society* 25, 149–174.

Pietarinen, Ahti-Veikko. Forthcoming. Which philosophy of mathematics is pragmaticism? In Moore (forthcoming 1).

Potter, Vincent. 1996. Peirce on continuity. In *Peirce's Philosophical Perspectives*, 117–123. Edited by Vincent M. Colapietro. New York: Fordham University Press.

Putnam, Hilary. 1992. Peirce's continuum. In Peirce (1992), 37–54.

Pycior, Helena M. 1995. Peirce at the intersection of mathematics and philosophy: A response to Eisele. In Ketner (1995), 132–145.

Rosenthal, Sandra. 1984. Mathematical necessity, scientific fallibilism and pragmatic verificationism: The Peircean synthesis. *International Philosophical Quarterly* (March), 3–20.

Sfendoni-Mentzou, Demetra. 1997. Peirce on continuity and laws of nature. *Transactions of the Charles S. Peirce Society* 33, 646–678.

Shields, Paul. 1997. Peirce's axiomatization of arithmetic. In Houser, Roberts, and Van Evra (1997), 43–52.

Shin, Sun-Joo. 1997. Kant's syntheticity revisited by Peirce. *Synthese* 113, 1–41.

———. Forthcoming. Peirce's two ways of abstraction. In Moore (forthcoming 1).

Stroh, Guy W. 1985. A note on Feibelman's interpretation of Peirce's conception of mathematics. *Transactions of the Charles S. Peirce Society* 21, 419–424.

Tiercelin, Claudine. 1993. Peirce's realistic approach to mathematics: Or, can one be a realist without being a platonist? In Moore (1993), 30–48.

———. Forthcoming. Peirce on mathematical objects and mathematical objectivity. In Moore (forthcoming 1).

Van Evra, James. 1997. Logic and mathematics in Charles S. Peirce's "Description of a notation for the logic of relatives." In Houser, Roberts, and Van Evra (1997), 147–157.

Zalamea, Fernando. 2001. *El continuo peirceano*. Bogotá: Universidad Nacional de Colombia.

———. 2003. Peirce's logic of continuity. *Review of Modern Logic* 9, 115–162.

———. Forthcoming. A category theoretic reading of Peirce's system: Pragmaticism, continuity and the existential graphs. In Moore (forthcoming 1).

Zeman, J. Jay. 1968. Peirce's graphs: The continuity interpretation. *Transactions of the Charles S. Peirce Society* 4, 144–154.

———. 1983. Peirce on abstraction. In Freeman (1983), 293–311.

3. OTHER WORKS CITED

Abbot, Francis E. 1885. *Scientific Theism.* Boston: Little, Brown.

Ammonius Hermiae. 1891. *In Porphyrii Isagogen sive V Voces.* Edited by Adolf Busse. *Commentaria in Aristotelem Graeca,* IV.3. Berlin: Reimer.

Armstrong, D. M. 1989. *Universals.* Boulder: Westview Press.

Baez, John C. 2002. The octonions. *Bulletin of the American Mathematical Society* 39, 145–205.

Bell, John L. 1998. *A Primer of Infinitesimal Analysis.* Cambridge: Cambridge University Press.

———. 2005. *The Continuous and the Infinitesimal in Mathematics and Philosophy.* Milan: Polimetrica.

Benacerraf, Paul. 1973. Mathematical truth. *Journal of Philosophy* 70, 661–679.

———. 1983. What numbers could not be. In P. Benacerraf and H. Putnam, eds., *Philosophy of Mathematics: Selected Readings, 2d ed.,* 272–294. Cambridge: Cambridge University Press.

Boler, John F. 1963. *Charles Peirce and Scholastic Realism.* Seattle: University of Washington Press.

———. 2004. Peirce and medieval thought. In Misak (2004), 58–86.

Boole, George. 1854. *An Investigation of the Laws of Thought.* London: Walton and Maberly. [Dover reprint edition, New York, 1951]

Borel, Émile. 1898. *Leçons sur la théorie des fonctions.* Paris: Gauthier-Villars.

Brady, Geraldine. 2000. *From Peirce to Skolem.* Amsterdam: North-Holland.

Brent, Joseph. 1993. *Charles Sanders Peirce.* Bloomington: Indiana University Press.

Burgess, John, and Gideon Rosen. 1997. *A Subject with No Object.* Oxford: Clarendon Press.

Cantor, Georg. 1872. Über die Ausdehnung eines Satzes der Theorie der trigonometrischen Reihen. *Mathematische Annalen* 5:122–132. [Cantor 1960, 92–101]

———. 1874. Über eine Eigenschaft des Inbegriffes aller reelen albegraischen Zahlen. *Journal für die reine und angewandte Mathematik* 77, 258–262.

———. 1878. Ein Beitrag zur Mannigfaltigkeitslehre. *Journal für die reine und angewandte Mathematik* 84, 242–258.

———. 1883. *Grundlagen einer allgemeinen Mannigfaltigkeitslehre.* Leipzig: B. G. Teubner. [English translation by William Ewald in Ewald (1996), 881–920]

———. 1885. Über verschiedene Theoreme aus der Theorie der Punktmengen in einem *n*-fach ausgedehnten stetigen Raume G_n. (Zweite Mitteilung.) *Acta Mathematica* 7, 105–124. [Cantor 1960, 261–277]

———. 1887–1888. Mitteilungen zur Lehre vom Transfiniten. *Zeitschrift für Philosophie und philosophische Kritik* 91 (1887), 81–125; 92 (1888), 240–265. [Cantor 1960, 378–439]

———. 1891. Über eine elementare Frage der Mannigfaltigkeitslehre. *Jahresbericht der Deutschen Mathematiker-Vereinigung* I:75–78.

———. 1895. Beiträge zur Begründung der transfiniten Mengenlehre, Part I. *Mathematische Annalen* 46:481–512. [English translation by P. E. B. Jourdain in Cantor (1915), 85–136]

———. 1897. Beiträge zur Begründung der transfiniten Mengenlehre, Part II. *Mathematische Annalen* 49:207–246.

———. 1915. *Contributions to the Founding of the Theory of Transfinite Numbers.* Translated by P. E. B. Jourdain. Chicago: Open Court. [Dover reprint, New York, 1955]

———. 1960. *Gesammelte Abhandlungen.* Hildesheim: Olm. Edited by Ernst Zermelo.

Cayley, Arthur. 1859. A sixth memoir upon quantics. *Philosophical Transactions of the Royal Society of London* 149, 61–90. [*The Collected Mathematical Papers of Arthur Cayley,* Vol. II, 561–592. Cambridge: Cambridge University Press, 1889]

Chessin, Alexander S. 1896. Demonstration of the existence of a limit for regular sequences of rational numbers. *Johns Hopkins University Circulars* 15(123), 37–38.

Chrystal, George. 1883. Mathematics. In *Encyclopaedia Britannica, 9th ed.* Edinburgh: Adam and Charles Black.

Clifford, William K. 1865–1866. Analytical Metrics. *Quarterly Journal of Pure and Applied Mathematics* 25 (February 1865), 54–67; 29 (June 1866), 16–21; 30 (October 1866), 119–126. [*Collected Mathematical Papers,* 80–109. Edited by Robert Tucker. London: Macmillan, 1882]

———. 1886. Philosophy of the pure sciences. In *Lectures and Essays,* 2d ed., 180–243. Edited by Leslie Stephen and Frederick Pollock.

Comte, Auguste. 1830–1842. *Cours de philosophie positive.* Paris: Bachelier. [*Introduction to Positive Philosophy.* Edited and translated, with introduction, by Frederick Ferré. Indianapolis: Hackett, 1966]

Dauben, Joseph W. 1990. *Georg Cantor: His Mathematics and Philosophy of the Infinite.* Princeton: Princeton University Press.

Dedekind, Richard. 1872. *Stetigkeit und irrationale Zahlen.* Braunschweig: Vieweg. [English translation by W. W. Beman, with corrections by W. B. Ewald, in Ewald (1996), 766–779]

———. 1888. *Was sind und was sollen die Zahlen?* Vieweg: Braunschweig. [English translation by W. W. Beman, in Ewald (1996), 790–833]

De Morgan, Augustus. 1847. *Formal Logic.* London: Taylor and Walton.

———. 1858. On the syllogism, III; and on logic in general. *Transactions of the Cambridge Philosophical Society* 10, 173–230. [De Morgan 1966, 74–146]

———. 1860a. Logic. In *The English Cyclopaedia (Arts and Sciences),* V, 340–354. Edited by Charles Knight. London: Bradbury and Evans. [De Morgan 1966, 247–270]

———. 1860b. On the syllogism IV; and on the logic of relations. *Transactions of the Cambridge Philosophical Society* 10, 331–358. [De Morgan 1966, 208–246]

———. 1860c. *Syllabus of a Proposed System of Formal Logic.* London: Walton and Maberly. [De Morgan 1966, 147–207]

———. 1863. On the syllogism V; and on various points of the onymatic system. *Transactions of the Cambridge Philosophical Society* 10, 428–487. [De Morgan 1966, 271–345]

———. 1966. *On the Syllogism and Other Logical Writings.* Edited by Peter Heath. London: Routledge and Kegan Paul, 1966.

De Tienne, André. 2001. "Scientific Fallibilism": Peirce's forgotten lecture of 1893. *Peirce Project Newletter* 4, 4–5,11.

Dipert, Randall. R. 1995. Peirce's underestimated place in the history of logic: A response to Quine. In Ketner (1995), 32–58.

———. 2004. Peirce's deductive logic: Its development, influence and philosophical significance. In Misak (2004), 287–324.

Duhamel, Jean-Marie. 1856. *Éléments de calcul infinitesimal.* Paris: Mallet-Bachelier.

Ehrlich, Philip. 2006. The rise of non-Archimedean mathematics and the roots of a misconception I: The emergence of non-Archimedean systems of magnitudes. *Archive for History of Exact Sciences* 60(1), 1–121.

Euclid. 1926. *The Thirteen Books of the Elements.* 2d ed. 3 vols. Cambridge: Cambridge University Press. Translated with introduction and commentary by Thomas L. Heath. [Dover reprint edition, New York, 1956]

Ewald, William, ed. 1996. *From Kant to Hilbert: A Source Book in the Foundations of Mathematics,* Volume II. Oxford: Clarendon Press.

Fermat, Pierre de. 1891–1912. *Oeuvres de Fermat.* 4 vols. Edited by Paul Tannery and Charles Henry. Paris: Gauthier-Villars.

Field, Hartry. 1980. *Science without Numbers.* Oxford: Blackwell.

Fisch, Max H. 1967. Peirce's progress from nominalism toward realism. *Monist* 51, 159–178. [Fisch 1986, 184–200]

———. 1971. Peirce's Arisbe: The Greek influence in his later philosophy. *Transactions of the Charles S. Peirce Society* 7, 187–210. [Fisch 1986, 227–248]

———. 1986. *Peirce, Semeiotic and Pragmatism.* Edited by Kenneth L. Ketner and Christian J. W. Kloesel. Bloomington: Indiana University Press.

Flower, Elizabeth, and Murray G. Murphey. 1977. *A History of Philosophy in America, Volume II.* New York: Capricorn.

Fogelin, Robert J. 1976. Hamilton's theory of quantifying the predicate—A correction. *Philosophical Quarterly* 26, 352–353.

Freeman, Eugene, ed. 1983. *The Relevance of Charles Peirce.* La Salle: Hegeler Institute.

Frege, Gottlob. 1884. *Die Grundlagen der Arithmetik.* Breslau: Koebner. [English translation by J. L. Austin, rev. 2d ed. Evanston, Ill.: Northwestern University Press, 1968]

Garner, Lynn E. 1981. *An Outline of Projective Geometry.* Amsterdam: North-Holland.

Goodman, Nelson and W. V. Quine. 1947. Steps toward a constructive nominalism. *Journal of Symbolic Logic* 12, 105–122.

Grajewski, Maurice. 1944. *The Formal Distinction of Duns Scotus.* Washington: Catholic University of America Press.

Grattan-Guinness, Ivor. 1997. *The Norton History of the Mathematical Sciences.* New York: W.W. Norton and Co.

Graves, Robert P. 1882–1889. *Life of Sir William Rowan Hamilton.* 3 vols. Dublin: Hodges, Figgis and Co.

Guyer, Paul, and Allen W. Wood. 1998. Introduction to Immanuel Kant, *Critique of Pure Reason.* Cambridge: Cambridge University Press.

Hamilton, William. 1860. *Lectures on Logic.* Edited by Henry Mansel and John Veitch. Boston: Gould and Lincoln.

Hamilton, William R. 1837. Theory of conjugate functions, or algebraic couples; with a preliminary and elementary essay on algebra as the science of pure time. *Transactions of the Royal Irish Academy* 17:293–422.

Hart, W. D., ed. 1996. *The Philosophy of Mathematics*. Oxford: Oxford University Press.

Hartshorne, Robin. 2000. *Geometry: Euclid and Beyond*. New York: Springer.

Heath, Thomas L. 1926. Introduction to Euclid (1926).

Hellman, Geoffrey. 1989. *Mathematics without Numbers*. Oxford: Clarendon.

Hilpinen, Risto. 1983. Peirce's theory of the proposition. In Freeman (1983), 264–270.

Hookway, Christopher. 1985. *Peirce*. London: Routledge.

Houser, Nathan. 1992. The fortunes and misfortunes of the Peirce papers. In Michael Balat, Janice Deledalle-Rhodes, Gérard Deledalle, and J. Deledalle, eds., *Signs of Humanity, Vol. 3*, 1259–1268. Berlin: Mouton de Gruyter.

Houser, Nathan, Don D. Roberts and James Van Evra, eds. 1997. *Studies in the Logic of Charles Sanders Peirce*. Bloomington: Indiana University Press.

Jevons, W. Stanley. 1864. *Pure Logic*. London: Edward Stanford.

Kempe, Alfred Bray. 1886. *A Memoir on the Theory of Mathematical Form*. London: Trübner.

Kent, Beverly. 1987. *Charles S. Peirce: Logic and the Classification of the Sciences*. Kingston and Montreal: McGill-Queen's University Press.

———. 1997. The interconnectedness of Peirce's diagrammatic thought. In Houser, Roberts, and Van Evra (1997), 445–459.

Ketner, Kenneth L., ed. 1995. *Peirce and Contemporary Thought*. New York: Fordham University Press.

Klein, Felix. 1871. Über die sogennante Nicht-Euklidische Geometrie. *Mathematische Annalen* 4, 573–625.

———. 1924–1948. *Elementarmathematik vom höheren Standpunkt aus*. 3 vols. Berlin, Springer. [English translation of Vol. 2 (*Geometry*) by E. R. Hedrick and C. A. Noble. New York: Macmillan, 1939.]

Kline, Morris. 1972. *Mathematical Thought from Ancient to Modern Times*. New York, Oxford University Press, 1972.

Kreisel, Georg. 1958. Wittgenstein's *Remarks on the Foundations of Mathematics*. *British Journal for the Philosophy of Science* 9, 135–158.

———. 1965. Mathematical logic. In T. L. Saaty, ed., *Lectures on Modern Mathematics*, Vol. III, 95–195. New York: Wiley.

Lange, Friedrich A. 1877. *Logische Studien*. Iserlohn: Baedeker.

Laudan, Larry. 1968. The *Vis viva* controversy, a post-mortem. *Isis* 59, 130–143.

Leibniz, Gottfried W. 1684. Meditations on knowledge, truth and ideas. [English translation in *Philosophical Essays*, 23–27. Edited and translated by Roger Ariew and Daniel Garber. Indianapolis: Hackett, 1989]

Listing, Johann B. 1847. Vorstudien zur Topologie. *Göttinger Studien*, 811–875.

———. 1862. Der Census räumlicher Complexe oder Verallgemeinerung des Euler'schen Satzes von den Polyëdern. *Abhandlungen der Königlichen Gesellschaft der Wissenschaften zu Göttingen* 10, 97–180.

Love, A. E. H. 1927. *A Treatise on the Mathematical Theory of Elasticity*, 4th ed. Cambridge: Cambridge University Press, 1927.

Maddy, Penelope. 1997. *Naturalism in Mathematics*. Oxford: Clarendon Press.

Markovic, Zeljko. 1970. Rudjer J. Boskovic. In Charles C. Gillespie, ed., *Dictionary of Scientific Biography*, Vol. II, 326–332. New York: Scribner's.

Masi, Michael. 1983. *Boethian Number Theory: A Translation of the* De Institutione Arithmetica. Amsterdam: Rodopi, 1983.

Maxwell, James Clerk. 1891. *A Treatise on Electricity and Magnetism*, 3d ed. 2 vols. [Dover reprint edition, New York, 1954]

Mayorga, Rosa. 2007. *From Realism to "Realicism."* Lanham: Lexington Books.

Meinong, Alexius. 1904. Über Gegenstandstheorie. In Alexius Meinong, ed., *Untersuchungen zur Gegenstandstheorie und Psychologie*, 1–50. Leipzig: Barth. [English translation by Isaac Levi, D. B. Terrell, and Roderick M. Chisholm, in Roderick M. Chisholm, ed., *Realism and the Background of Phenomenology*, 76–117. Glencoe: Free Press, 1960]

Mill, John Stuart. 1843. *A System of Logic*. London: J. W. Parker. [Mill 1963–1991, vols. 7–8.]

———. 1865. *An Examination of Sir William Hamilton's Philosophy*. London: Longman, Green. [Mill 1963–1991, vol. 9, incorporating addenda and variants from later editions]

———. 1963–1991. *Collected Works*. 33 vols. Edited by John M. Robson. Toronto: University of Toronto Press.

Misak, Cheryl, ed. 2004. *The Cambridge Companion to Peirce*. Cambridge: Cambridge University Press.

Möbius, August F. 1827. *Der barycentrische calcul*. Leipzig: J.A. Barth.

Moore, A. W. 1990. *The Infinite*. London: Routledge.

Moore, Edward C., ed. 1993. *Charles S. Peirce and the Philosophy of Science.* Tuscaloosa: University of Alabama Press.

Moore, Gregory H. 1982. *Zermelo's Axiom of Choice.* New York: Springer.

Mueller, Ian. 1981. *Philosophy of Mathematics and Deductive Structure in Euclid's* Elements. Cambridge, Mass.: MIT Press.

Murphey, Murray G. 1961. *The Development of Peirce's Philosophy.* Cambridge, Mass.: Harvard University Press. [Hackett reprint edition, Indianapolis, 1993]

O'Hara, Charles W., and Dudley R. Ward. 1937. *Introduction to Projective Geometry.* Oxford: Clarendon Press.

Parsons, Charles. 1979–1980. Mathematical intuition. *Proceedings of the Aristotelian Society* 80, 145–168. [Hart 1996, 95–113]

———. 1990. The structuralist view of mathematical objects. *Synthese* 84, 303–346. [Hart 1996, 272–309]

Peirce, Benjamin. 1870. *Linear Associative Algebra.* Washington City.

Peterson, Sven R. 1955. Benjamin Peirce: Mathematician and philosopher. *Journal of the History of Ideas* 16, 89–112.

Plücker, Julius. 1839. *Theorie der algebraischen Kurven.* Bonn: Adolf Marcus.

Przytycki, Józef H. 1998. Classical roots of knot theory. *Chaos, Solitions and Fractals* 9, 531–545.

Putnam, Hilary. 1979. What is mathematical truth? In *Philosophical Papers, Vol. 1: Mathematics, Matter and Method,* 60–78. Cambridge: Cambridge University Press.

———. 1982. Peirce the logician. *Historia Mathematica* 9, 290–301.

Quine, W. V. O. 1969. Ontological relativity. In *Ontological Relativity & Other Essays,* 26–68. New York: Columbia University Press.

Resnik, Michael. 1997. *Mathematics as a Science of Patterns.* Oxford: Clarendon Press.

Riemann, Georg F. B. 1851. Grundlagen für eine allgemeine Theorie der Functionen einer verändlichen complexen Grösse. In *Gesammelte Mathematische Werke und wissenschaftlichen Nachlass,* 2d ed., 3–45. Edited by Heinrich Weber and Richard Dedekind. Leipzig: Teubner, 1892.

———. 1867. Über die Hypothesen, welche der Geometrie zu Grunde liegen. *Abhandlungen der Königlichen Gesellschaft der Wissenschaften zu Göttingen* 13:133–152.

Roberts, Don D. 1973. *The Existential Graphs of Charles S. Peirce.* The Hague: Mouton.

Robin, Richard. 1967. *Annotated Catalogue of the Papers of Charles S. Peirce.* Amherst: University of Massachusetts.

Robinson, Abraham. 1966. Non-standard analysis. Princeton: Princeton University Press. [Second edition, Princeton: Princeton University Press, 1973]

Royce, Josiah. 1899. *The World and the Individual, First Series: The Four Historical Conceptions of Being.* New York: Macmillan.

Russell, Bertrand. 1902. *Principles of Mathematics.* Cambridge: Cambridge University Press. [2d ed., reprint, New York: Norton, 1938]

———. 1957. Mathematics and the metaphysicians. In *Mysticism and Logic,* 70–92. Garden City: Doubleday.

Schröder, Ernst. 1880–1905. *Vorlesungen über die Algebra der Logik.* 3 vols. Leipzig: Teubner.

Schubring, Gert. 2005. *Conflicts between Generalization, Rigor and Intuition.* New York: Springer.

Shapiro, Stewart. 1997. *Philosophy of Mathematics: Structure and Ontology.* Oxford: Oxford University Press.

———. 2000. *Thinking about Mathematics.* Oxford: Oxford University Press.

Shin, Sun-Joo. 2002. *The Iconic Logic of Peirce's Graphs.* Cambridge, Mass.: MIT Press.

Short, Thomas L. 1988. Hypostatic abstraction in empirical science. *Grazer Philosophische Studien* 32, 51–68.

———. 2007. *Peirce's Theory of Signs.* Cambridge: Cambridge University Press.

Skyrms, Brian. 2000. *Choice and Chance.* 4th ed. Belmont: Wadsworth.

Staudt, Karl G. C. Von. 1847. *Geometrie der Lage.* Nürnberg: Korn.

Stewart, Ian, and David Tall. 1983. *Complex Analysis.* Cambridge: Cambridge University Press.

Stjernfelt, Frederik. 2000. Diagrams as centerpiece of a Peircean epistemology. *Transactions of the Charles S. Peirce Society* 36, 357–392.

Thomson, William and Peter G. Tait. 1867. *Treatise on Natural Philosophy.* Oxford: Clarendon Press.

Tiles, J. E. 1988. Iconic thought and the scientific imagination. *Transactions of the Charles S. Peirce Society* 24, 161–178.

Torretti, Roberto. 1978. *Philosophy of Geometry from Riemann to Poincaré.* Dordrecht: Reidel.

Weierstrass, Karl. 1895. Über continuirliche Functionen eines reelen Arguments, die für keinen Werth des letzteren einen bestimmten Differentialquotienten besitzen. In *Mathematische Werke,* Vol. 2, 71–74. Berlin: Mayer and Müller.

Weiss, Paul. 1934. Charles Sanders Peirce. In Dumas Malone, ed., *Dictionary of American Biography,* 398–403. New York: Scribner's.

Whitney, William Dwight, ed. 1889. *The Century Dictionary: An Encyclopedic Lexicon of the English Language.* New York: The Century Co.

———, ed. 1909. *The Century Dictionary Supplement.* New York: The Century Co.

Wilson, Leonard. 1973. Uniformitarianism and catastrophism. In Philip P. Wiener, ed., *Dictionary of the History of Ideas,* Vol. IV, 417–423. New York: Scribner's.

Wittgenstein, Ludwig. 1953. *Philosophical Investigations.* Translated by G. E. M. Anscombe. New York: Macmillan. [3d ed., 1968]

Zermelo, Ernst. 1904. Beweis, daß jede Menge wohlgeordnet werden kann. *Mathematische Annalen* 59, 514–516.

Index